Deflection of Beams for All Spans and Cross Sections

Deflection of Beams for All Spans and Cross Sections

Yun C. Ku

McGraw-Hill Book Company

New York   St. Louis   San Francisco   Auckland   Bogotá
Hamburg   Johannesburg   London   Madrid   Mexico
Montreal   New Delhi   Panama   Paris
São Paulo   Singapore   Sydney
Tokyo   Toronto

Library of Congress Cataloging in Publication Data

Ku, Yun C.
  Deflection of beams for all spans and cross sections.

  Includes index.
  1. Girders. 2. Flexure. I. Title.
TA660.B4K77 1986   624.1'7723   84-29694
ISBN 0-07-035603-3

Copyright © 1986 by McGraw-Hill, Inc. All rights reserved.
Printed in the United States of America. Except as permitted
under the United States Copyright Act of 1976, no part of
this publication may be reproduced or distributed in any form
or by any means, or stored in a data base or retrieval
system, without prior written permission of the publisher.

1 2 3 4 5 6 7 8 9 0   DOC/DOC   8 9 8 7 6 5

ISBN 0-07-035603-3

The editors for this book were Joan Zseleczky and William B.
O'Neal, the designer was Elliot Epstein, and the production
supervisor was Sally L. Fliess. It was set in Caledonia
by Techna Type.

Printed and bound by R. R. Donnelley & Sons.

Contents

Preface   vii

1   Techniques for Computing the Deflection of Beams   1
1-1   Modulus of Elasticity and Moment of Inertia, 1
1-2   Deflection of Statically Determinate Structures, 6
1-3   Moment-Area Method, 7
1-4   Conjugate-Beam Method, 10
1-5   The Limits of Deflections, 12

2   Deflection of Simple-Support Beams   15
2-1   Simple-Support Beam with Concentrated Load, 15
2-2   Maximum Deflection of Concentrated Load, 17
2-3   Simple-Support Beam with Uniformly Distributed Load, 22
2-4   Maximum Deflection of Uniformly Distributed Load, 25

3   Deflection of Beams Other Than Simple-Support Beams   37
3-1   Effect of End Moments, 37
3-2   Deflection of Beams Due to End Moments, 40
3-3   Deflection of a Beam, Fixed at One End and Supported at the Other, 42
3-4   Deflection of a Beam, Fixed at Both Ends, 50
3-5   Combining the Equations of Deflection for Loads and End Moments, 58

4   Deflection of Continuous Beams   81
4-1   Support Moments, 81
4-2   Three-Moment Equation Method, 81
4-3   Consistent Deflection Method, 84
4-4   Moment-Distribution Method, 87
4-5   Influence Lines for Deflection of Continuous Beams, 90

5   Deflection of Cover-Plated Beams   119
5-1   Effect of Cover Plates, 119
5-2   Cover-Plated Beam with Concentrated Load, 121
5-3   Cover-Plated Beam with Uniformly Distributed Load, 127
5-4   Cover-Plated Beam with End Moments, 132

6 Deflection of Cover-Plated Continuous Beams   147

- 6-1 Cover Plates on Continuous Beams, 147
- 6-2 Fixed-End Moments, Stiffnesses, and Carryover Factors, 147
- 6-3 Application of Moment Distribution for Cover-Plated, Continuous Beams, 157

7 Deflection of Trusses   165

- 7-1 Introduction, 165
- 7-2 Unit-Load Method, 165
- 7-3 Maximum Deflection of Trusses, 167
- 7-4 Influence Lines, 169
- 7-5 Indeterminate Trusses, 175

Appendixes

- A Beam Diagrams and Formulas for Various Static Loading Conditions, 179
- B Charts for Determining Stiffness Factors, Carryover Factors, and Fixed-End Moments for Beams with Variable Cross Sections, 199
- C Units of Measure and Conversion, 219

Index   221

Preface

In the design of structures, the deflection of beams is always an important concern for the engineer. It indicates stiffness, rigidity, and safety of a structural unit. In certain cases, computation must be done to see that the deflection of the structure does not exceed the specified limits, in order to be sure that the design is safe. Perhaps the most important reason for an engineer's interest in deflection computation is that the stress analysis of statically indeterminate structures is based largely on an evaluation of their deflection under load. Because the computation is sometimes very laborious, especially for beams with variable cross sections, simplifying the solution of these problems is necessary.

Tables and equations for the solution of the coefficient of deflection are often used to aid the calculations. But using tables and equations blindly is dangerous. It is necessary to understand how basic principles of mechanics lead to these simplified solutions. This book contains solutions for the deflection of beams including simple beams, continuous beams, and beams reinforced with cover plates at the center or at the end of the beams.

We know that when a load is applied to a beam, it will deflect and produce a certain amount of bending moment in the beam. The bending moment is in some ways a proportional deflection. We also know that some materials are hard to deflect, while others are easier to deflect. How effectively a member can resist deflection will depend on the magnitude of the section modulus and the modulus of elasticity of the material. The greater the section modulus and modulus of elasticity of the material, the harder it is to deflect.

Furthermore, the beam will undergo less deflection if the end of the beam is restrained, or partially restrained, or reinforced by cover plates. All these conditions will produce upward deflection against the downward deflection due to the loading. In highway bridge stringer design, reduction in moment by using continuous supports and by placing cover plates on the maximum moment region produces the most efficient solution.

Basically speaking, the deflection of a continuous beam, carrying a uniformly distributed load or concentrated loads, is a combination of deflection due to the load between the supports and deflection due to the end moments at the supports. A general equation, therefore, can be derived on the basis of this assumption. On the other hand, we can also use influence lines to calculate the deflection of a continuous beam if the cross section of the beam is constant. For beams with variable cross sections, equations and tables can be used to aid the solution.

The author is very grateful to the Iowa Engineering Experiment Station for its permission to reproduce a series of 36 charts for determining the stiffness factors, carryover factors, and fixed-end moments for beams in which there are abrupt changes in the cross sections. These charts have been printed along with their derivations in Bulletin 176 of the Iowa Engineering Experiment Station. Using these charts, the engineer can readily find the necessary values and obtain the final end moments.

Thanks are also extended to the American Institute of Steel Construction, Inc., for its permission to reproduce tables of beam diagrams and formulas from the *Manual of Steel Construction*, frequently used for structural designing. However, to comply with the book's organization, the author rearranged the formulas; also, some have been deleted and others added as required.

Special thanks go to Dr. William H. Ku, Professor of Physics, at Columbia University, New York, for his help in generating the tables that aid in the solution of the deflection problems.

February 1984 New York, New York *Yun C. Ku*

Deflection of Beams for All Spans and Cross Sections

1
Techniques for Computing the Deflection of Beams

1-1 MODULUS OF ELASTICITY AND MOMENT OF INERTIA

Since we know that the modulus of elasticity E and the moment of inertia I play important roles in deflection calculations, it is necessary to know how these two factors affect deflections.

When a material under stress does not suffer permanent deformation upon removal of the stress, it is called *elastic*. Many materials are elastic up to a certain unit stress. Within the elastic limit, the ratio of the unit stress to the corresponding unit strain is constant and is called the *modulus of elasticity*, or *Young's modulus*. This simple relation is known as *Hooke's law*, expressed by equation as $f_x = E\epsilon_x$, where f_x is the normal stress in the x direction, ϵ_x is the normal strain in the same direction, and E is the modulus of elasticity.

A higher-strength material will have a higher modulus of elasticity and a higher degree of stiffness. The value of E for steel varies with the grade and generally ranges from 29,000 to 31,000 kilopounds per square inch (kips/in^2); for concrete it ranges from 2,000 to 6,000 kips/in^2; and for timber it varies with the species and to some extent the moisture content. For hard woods such as oak and beech 1,700 kips/in^2 is a fair average value, while softwoods have a lower value of E, approximately 1,350 kips/in^2. Therefore, if wood and steel have the same cross section, the deflection of the wood could be as much as 20 times greater than that of steel under the same load and for the same span length.

The concept of the *moment of inertia I* is closely linked with the flexure formula. The moment of inertia may be defined as the sum of the products obtained by multiplying all the elementary areas of a cross section by the squares of their distances from a given axis, i.e., $I = \int y^2 \, dA$. It may be found for any axis, but the axis commonly used is the neutral axis of a beam cross section, which, as we know, passes through the centroid of the section. The unit of measure for the moment of inertia is in^4.

We may note from the expression $\int y^2 \, dA$ that I varies with the shape of the cross section as well as with the area. It is, of course, a matter of common

TABLE 1-1 MOMENTS OF INERTIA AND SECTION MODULI OF VARIOUS CROSS SECTIONS

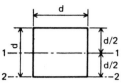

$A = d^2$
$I_1 = \dfrac{d^4}{12}$ $S_1 = \dfrac{d^3}{6}$
$I_2 = \dfrac{d^4}{3}$ $S_2 = \dfrac{d^3}{3}$
$r_1 = 0.2887d$
$r_2 = 0.5774d$

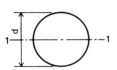

$A = \dfrac{\pi d^2}{4} = 0.7854d^2$
$I_1 = \dfrac{\pi d^4}{64} = 0.0491d^4$
$S_1 = \dfrac{\pi d^3}{32} = 0.0982d^3$
$r_1 = \dfrac{d}{4}$

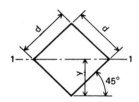

$A = d^2$
$y = 0.7071d$
$I_1 = \dfrac{d^4}{12}$ $S_1 = 0.1178d^3$
$r_1 = 0.2887d$

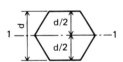

$A = 0.866d^2$
$I_1 = 0.06d^4$
$S_1 = 0.12d^3$
$r_1 = 0.264d$

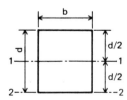

$A = bd$
$I_1 = \dfrac{bd^3}{12}$ $S_1 = \dfrac{bd^2}{6}$
$I_2 = \dfrac{bd^3}{3}$ $S_2 = \dfrac{bd^2}{3}$
$r_1 = 0.2887d$
$r_2 = 0.5774d$

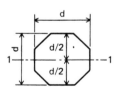

$A = 0.8284d^2$
$I_1 = 0.0547d^4$
$S_1 = 0.1094d^3$
$r_1 = 0.257d$

Note: A = area of section, I = moment of inertia, S = section modulus, r = radius of gyration = $\sqrt{I/A}$.

TABLE 1-1 CONTINUED

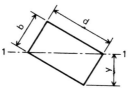

$A = bd$

$y = \dfrac{bd}{\sqrt{b^2 + d^2}}$

$I_1 = \dfrac{b^3 d^3}{6(b^2 + d^2)}$

$S_1 = \dfrac{b^2 d^2}{6\sqrt{(b^2 + d^2)}}$

$r_1 = \dfrac{bd}{\sqrt{6(b^2 + d^2)}}$

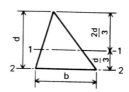

$A = \dfrac{bd}{2}$

$I_1 = \dfrac{bd^3}{36} \qquad S_1 = \dfrac{bd^2}{24}$

$I_2 = \dfrac{bd^3}{12} \qquad S_2 = \dfrac{bd^2}{12}$

$r_1 = 0.2357d$

$r_2 = 0.4082d$

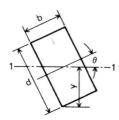

$A = bd$

$y = \dfrac{b \sin \theta + d \cos \theta}{2}$

$I_1 = \dfrac{bd(b^2 \sin^2 \theta + d^2 \cos^2 \theta)}{12}$

$S_1 = \dfrac{bd(b^2 \sin^2 \theta + d^2 \cos^2 \theta)}{6(b \sin \theta + d \cos \theta)}$

$r_1 = \sqrt{\dfrac{b^2 \sin^2 \theta + d^2 \cos^2 \theta}{12}}$

$A = \dfrac{d}{2}(b + b')$

$y_1 = \dfrac{d(2b + b')}{3(b + b')}$

$y = \dfrac{d(b + 2b')}{3(b + b')}$

$I_1 = \dfrac{d^3(b^2 + 4bb' + b'^2)}{36(b + b')}$

$S_1 = \dfrac{d^2(b^2 + 4bb' + b'^2)}{12(2b + b')}$

$r_1 = \dfrac{d}{6(b + b')}\sqrt{2(b^2 + 4bb' + b'^2)}$

Note: A = area of section, I = moment of inertia, S = section modulus, r = radius of gyration = $\sqrt{I/A}$.

TABLE 1-1 CONTINUED

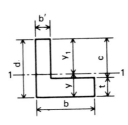

$A = bt + b'c$

$y = \dfrac{d^2b' + t^2(b - b')}{2(bt + b'c)}$

$y_1 = d - y$

$I_1 = \dfrac{b'y_1^3 + by^3 - (b - b')(y - t)^3}{3}$

$r_1 = \sqrt{\dfrac{I_1}{A}}$

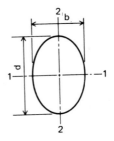

Ellipse

$A = 0.7854bd$

$I_1 = \dfrac{\pi bd^3}{64} = 0.0491bd^3$

$I_2 = \dfrac{\pi b^3 d}{64} = 0.0491b^3 d$

$S_1 = \dfrac{\pi bd^2}{32} = 0.0982bd^2$

$S_2 = \dfrac{\pi b^2 d}{32} = 0.0982b^2 d$

$r_1 = \dfrac{d}{4} \quad r_2 = \dfrac{b}{4}$

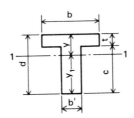

$A = bt + b'c$

$y = \dfrac{d^2b' + t^2(b - b')}{2(bt + b'c)}$

$y_1 = d - y$

$I_1 = \dfrac{b'y_1^3 + by^3 - (b - b')(y - t)^3}{3}$

$r_1 = \sqrt{\dfrac{I_1}{A}}$

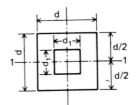

$A = d^2 - d_1^2$

$I_1 = \dfrac{d^4 - d_1^4}{12}$

$S_1 = \dfrac{d^4 - d_1^4}{6d}$

$r_1 = \sqrt{\dfrac{d^2 + d_1^2}{12}}$

Note: A = area of section, I = moment of inertia, S = section modulus, r = radius of gyration = $\sqrt{I/A}$.

TABLE 1-1 CONTINUED

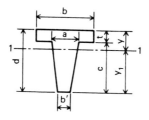

$A = bt + \dfrac{c(a + b')}{2}$

$y = $
$\dfrac{3bt^2 + 3b'c(d + t) + c(a - b')(c + 3t)}{3[2bt + c(a + b')]}$

$y_1 = d - y$

$I_1 = \dfrac{4bt^3 + c^3(3b' + a)}{12} - A(y - t)^2$

$r_1 = \sqrt{\dfrac{I_1}{A}}$

$A = bd - b_1 d_1$

$I_1 = \dfrac{bd^3 - b_1 d_1^3}{12}$

$S_1 = \dfrac{bd^3 - b_1 d_1^3}{6d}$

$r_1 = \sqrt{\dfrac{bd^3 - b_1 d_1^3}{12A}}$

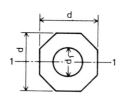

$A = 0.8284d^2 - 0.7854d_1^2$

$I_1 = 0.0547d^4 - 0.0491d^4$
$ = 0.0491(1.115d^4 - d_1^4)$

$S_1 = \dfrac{0.0982(1.115d^4 - d_1^4)}{d}$

$r_1 = \frac{1}{4}\sqrt{1.056d^2 + d_1^2}$

$A = \dfrac{\pi(d^2 - d_1^2)}{4} = 0.7854(d^2 - d_1^2)$

$I_1 = \dfrac{\pi(d^4 - d_1^4)}{64} = 0.0491(d^4 - d_1^4)$

$S_1 = \dfrac{\pi(d^4 - d_1^4)}{32d} = 0.0982\dfrac{d^4 - d_1^4}{d}$

$r_1 = \dfrac{1}{4}\sqrt{d^2 + d_1^2}$

Note: A = area of section, I = moment of inertia, S = section modulus, r = radius of gyration = $\sqrt{I/A}$.

TABLE 1-1 CONTINUED

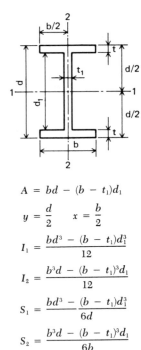

$A = bd - (b - t_1)d_1$

$y = \dfrac{d}{2} \qquad x = \dfrac{b}{2}$

$I_1 = \dfrac{bd^3 - (b - t_1)d_1^3}{12}$

$I_2 = \dfrac{b^3d - (b - t_1)^3 d_1}{12}$

$S_1 = \dfrac{bd^3 - (b - t_1)d_1^3}{6d}$

$S_2 = \dfrac{b^3d - (b - t_1)^3 d_1}{6b}$

$r_1 = \sqrt{\dfrac{I_1}{A}} \qquad r_2 = \sqrt{\dfrac{I_2}{A}}$

$A = bd - (b - t_1)d_1$

$I_3 = I_1 \sin^2 \theta + I_2 \cos^2 \theta$

$I_4 = I_1 \cos^2 \theta + I_2 \sin^2 \theta$

$f_b = M \left(\dfrac{d}{2I_1} \sin \theta + \dfrac{b}{2I_2} \cos \theta \right)$

where f_b = bending stress

M = bending moment due to force P

Note: A = area of section, I = moment of inertia, S = section modulus, r = radius of gyration = $\sqrt{I/A}$.

observation that a wood plank with a cross section of, say, 2 in by 6 in is much stiffer when used on a given span with its 6-in side vertical than when the same plank is placed with its 2-in side vertical.

The values of I for structural steel shapes usually can be found from tables in books on structural steel. Table 1-1 shows moments of inertia and section moduli for various cross sections.

1-2 DEFLECTION OF STATICALLY DETERMINATE STRUCTURES

There are several methods for finding the deflections of statically determinate structures, such as the unit-load method, Castigliano's theorem of least work, the moment-area method, and the conjugate-beam method. Perhaps the two methods most commonly used and applicable to both beams and frames are the unit-load method and the moment-area method. Castigliano's theorem of least work actually involves processes identical to those of the unit-load method, but the approach of the unit-load method is simpler. The conjugate-beam method is similar to the moment-area method. It can be considered an extension of the moment-area method. However, both the unit-load method and Castigliano's theorem of least work can compute only one particular deflection component at one time. The moment-area method and the conjugate-beam method can compute several deflection components simultaneously from which the influence line of the deflection of the beam can be obtained. Therefore, these two methods are the most effective ones for computing the deflection of beams.

3 MOMENT-AREA METHOD

The moment-area method has two theorems:

Theorem I: The angle between the tangents to the elastic curve through any two points of the axis of the beam under bending is equal to the area of the M/EI diagram between these two points.

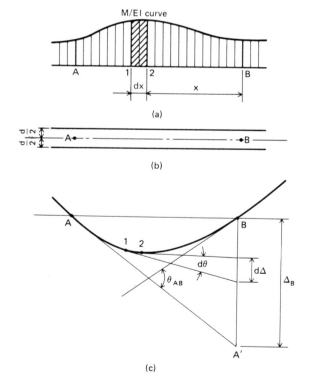

Fig. 1-1. (a) M/EI diagram; (b) member AB; (c) elastic curve.

8 Chapter 1

Theorem II: The deflection at any point on the axis of the beam under bending, measured from the tangent through any other point, equals the moment about the first point of the M/EI diagram between the two points.

For example, Fig. 1-1 shows an M/EI curve on a straight member AB. The area of the curve between A and B is the moment area of M/EI between those two points. If the angle θ_{AB} denotes the change of slope between the tangents to the elastic curve through points A and B of the member AB, and EI is constant, then

$$\theta_{AB} = \int_A^B \frac{M \, dx}{EI} \tag{1-1}$$

and Δ_B, the deflection of the point B of a member AB from the tangent at A, is given by

$$\Delta_B = \int_A^B \frac{Mx \, dx}{EI} \tag{1-2}$$

Therefore, in applying the moment-area method to find the deflections and rotations of a beam, it is necessary to plot or sketch the moment diagram (or the M/I diagram, if I is variable) first, and then to draw a qualitative picture of the elastic curve.

Example 1-1

Find θ_B and Δ_B by the moment-area method for a cantilever beam loaded with a concentrated load P at the end B as shown in Fig. 1-2.

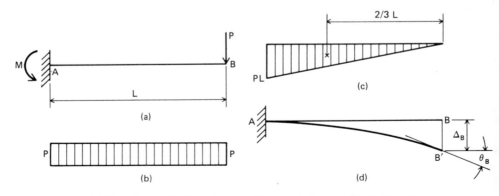

Fig. 1-2. (a) Given beam; (b) shear diagram; (c) moment diagram; (d) elastic curve.

Solution

By Eq. (1-1)

$$\theta_B = \theta_A + \text{area of the } M/EI \text{ diagram between } A \text{ and } B$$

$$EI\theta_B = 0 + \frac{(PL)(L)}{2} = \frac{PL^2}{2}$$

$$\theta_B = \frac{PL^2}{2EI} \quad \text{clockwise} \qquad (ans.) \qquad (1\text{-}3)$$

By Eq. (1-2)

$$\Delta_B = \text{moment of the } M/EI \text{ diagram between } A \text{ and } B \text{ about the deflection of } B$$

$$EI\Delta_B = \frac{PL^2}{2} \frac{2L}{3} = \frac{PL^3}{3}$$

$$\Delta_B = \frac{PL^3}{3EI} \quad \text{in downward}^1 \qquad (ans.) \qquad (1\text{-}4)$$

Example 1-2

Find θ_B and Δ_B by the moment-area method for a cantilever beam loaded with a uniformly distributed load as shown in Fig. 1-3.

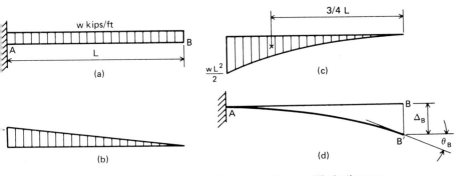

Fig. 1-3. (a) Given beam; (b) shear diagram; (c) moment diagram; (d) elastic curve.

Solution

By Eq. (1-1)

$$\theta_B = \theta_A + \text{area of the } M/EI \text{ diagram between } A \text{ and } B$$

$$EI\theta_B = 0 + \frac{1}{3}\frac{wL^2}{2}L = \frac{wL^3}{6}$$

$$\theta_B = \frac{wL^3}{6EI} \quad \text{clockwise} \qquad (ans.) \qquad (1\text{-}5)$$

---
[1] Units of measure for P are kips, kips/in² for E, in⁴ for I, ft × 12 in for L, and in for Δ (see Appendix C for metric conversion).

By Eq. (1-2)

$$\Delta_B = \text{moment of the } M/EI \text{ diagram between } A \text{ and } B \text{ about the deflection of } B$$

$$EI\Delta_B = \frac{wL^3}{6}\frac{3L}{4} = \frac{wL^4}{8}$$

$$\Delta_B = \frac{wL^4}{8EI} \quad \text{in downward}[2] \qquad (ans.) \qquad (1-6)$$

1-4 CONJUGATE-BEAM METHOD

In Examples 1-1 and 1-2 the elastic curve has a horizontal tangent at a certain point of the beam. The slope and the deflection at any other point can be found by referring to this horizontal tangent as a basis where the two moment-area theorems are applied. In many cases, however, there may be no horizontal tangent, such as for the beam shown in Fig. 1-4. It is then necessary to find the slope θ_C and the deflection Δ_C on the beam.

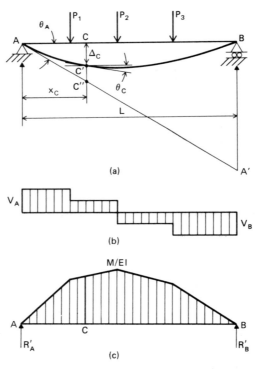

Fig. 1-4. (a) Given beam; (b) shear diagram; (c) moment diagram.

[2] wL is in kips.

To do this, draw a tangent line AA' from point A. If θ_A is the angle of slope at point A, then

$$\theta_C = \theta_A - \text{area of } \frac{M}{EI} \text{ diagram between } A \text{ and } C$$

where $\theta_A = \dfrac{A'B}{AB}$

$$= \frac{1}{L}\left(\text{moment of } \frac{M}{EI} \text{ diagram between } A \text{ and } B \text{ about } B\right)$$

Thus

$$\theta_C = \frac{1}{L}\left(\text{moment of } \frac{M}{EI} \text{ diagram between } A \text{ and } B \text{ about } B\right)$$

$$- \left(\text{area of } \frac{M}{EI} \text{ diagram between } A \text{ and } C\right) \quad (1\text{-}7)$$

Since $\Delta_C = CC'' - C'C''$

and $CC'' = \theta_A x_C$

$$= \frac{1}{L}\left(\text{moment of } \frac{M}{EI} \text{ diagram between } A \text{ and } B \text{ about } B\right) x_C$$

and $C'C'' = $ deflection of C from tangent at A

$$= \text{moment of } \frac{M}{EI} \text{ diagram between } A \text{ and } C \text{ about } C$$

then

$$\Delta_C = \frac{1}{L}\left(\text{moment of } \frac{M}{EI} \text{ diagram between } A \text{ and } B \text{ about } B\right) x_C$$

$$- \left(\text{moment of } \frac{M}{EI} \text{ diagram between } A \text{ and } C \text{ about } C\right) \quad (1\text{-}8)$$

Suppose that the beam AB is loaded by the M/EI diagram as shown in Fig. 1-4(c). The slope and the deflection of the beam can be obtained by this imaginary load. This beam is called a *conjugate beam*. We assume that R'_A and R'_B represent the reactions of the conjugate beam and that V'_C and M'_C represent the shear and bending moment at C of the conjugate beam. The right side of Eq. (1-7) is obviously equal to V'_C, and the right side of Eq. (1-8) is obviously equal to M'_C. Thus

$$\theta_C = V'_C \quad (1\text{-}9)$$

$$\Delta_C = M'_C \quad (1\text{-}10)$$

where V'_C = shear of the M/EI diagram at point C

M'_C = moment of the M/EI diagram at point C

If we assume that the maximum deflection occurs at a point E, then $V'_E = 0$ and

$$\Delta_{\max} = M'_E \qquad (1\text{-}11)$$

From Eqs. (1-9) and (1-10), the conjugate-beam method may be stated as follows:

Theorem I: The angle between the chord AB and the tangent to the elastic curve at any point C between points A and B on the elastic curve is equal to the shear of the M/EI diagram at point C.

Theorem II: The deflection of any point C between two points A and B on the elastic curve, measured from the chord AB, is equal to the bending moment of the M/EI diagram at point C.

These two equations can be applied between any two points A and B on the elastic curve, unless the chord AB is not horizontal.

1-5 THE LIMITS OF DEFLECTIONS

The term "deflection" as used herein will refer to the deflection computed in accordance with design loads.

The *Manual of Steel Construction*,[3] published by the American Institute of Steel Construction, Inc. (AISC), requires for buildings that the live load deflection of floor beams supporting plastered ceilings be limited to not more than 1/360 of the span length.

The American Association of the State Highway and Transportation Officials (AASHTO), in its *Standard Specifications for Highway Bridges*,[4] requires that structural members having simple or continuous spans be designed so that the deflection due to live load plus impact does not exceed 1/800 of the span; on bridges in urban areas, used in part by pedestrians, the ratio preferably should be 1/1,000.

The deflection of cantilever arms due to live load plus impact should be limited to 1/300 of the cantilever arm, except for the case involving pedestrian use, where the ratio preferably should be 1/375.

When spans have cross bracings or diaphragms sufficient in depth or

---

[3] *Manual of Steel Construction*, 8th ed., American Institute of Steel Construction, 1980.

[4] *Standard Specifications for Highway Bridges*, 12th ed., American Association of the State Highway and Transportation Officials, 1977.

strength to ensure lateral distribution of loads, the deflection may be computed for the standard H or HS loading, considering all beams or stringers to be acting together and having equal deflection.

The maximum deflection due to live load and impact should be designed within the limit prescribed by the specifications or codes in order to be sure that the designed structural unit will be safe and comfortable in service.

2
Deflection of Simple-Support Beams

1 SIMPLE-SUPPORT BEAM WITH CONCENTRATED LOAD

For computing the deflection, a general equation has to be derived. Assume that a simple-support beam AB carries a concentrated load P at point C as shown in Fig. 2-1. We will find the deflection of the beam by the conjugate-beam method.

From Fig. 2-1(c), we obtain

$$\text{Area I} = \frac{Pa^2b}{2L} \quad \text{kip} \cdot \text{ft}^2$$

$$\text{Area II} = \frac{Pab^2}{2L} \quad \text{kip} \cdot \text{ft}^2$$

$$R'_A = \frac{Pab(a+2b)}{6L} \quad \text{kip} \cdot \text{ft}^2$$

$$R'_B = \frac{Pab(b+2a)}{6L} \quad \text{kip} \cdot \text{ft}^2$$

$$\theta_A = \frac{R'_A}{EI} = \frac{Pab}{6EIL}(a+2b) \quad \text{rad clockwise}$$

$$\theta_B = \frac{R'_B}{EI} = \frac{Pab}{6EIL}(b+2a) \quad \text{rad counterclockwise}$$

Considering the left portion of the beam, Fig. 2-1(d), when $x < a$

$$M'_x = \frac{PL^3}{6} \frac{x}{L}\left(1 - \frac{a}{L}\right)\left(2\frac{a}{L} - \frac{a^2}{L^2} - \frac{x^2}{L^2}\right) \quad \text{kip} \cdot \text{ft}^3$$

$$\Delta_x = \frac{PL^3}{6EI} \frac{x}{L}\left(1 - \frac{a}{L}\right)\left(2\frac{a}{L} - \frac{a^2}{L^2} - \frac{x^2}{L^2}\right) \quad (2\text{-}1)$$

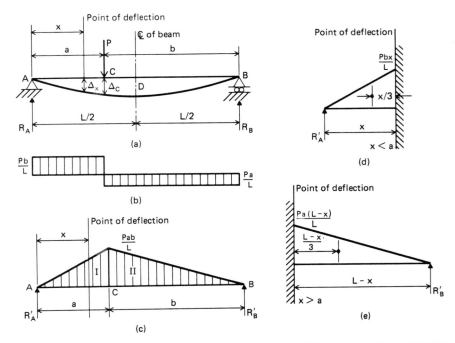

Fig. 2-1. (a) Given beam; (b) shear diagram; (c) moment diagram; (d) left portion of beam; (e) right portion of beam.

For the right portion of the beam, Fig. 2-1(e), when $x > a$

$$M'_x = \frac{PL^3}{6} \frac{a}{L}\left(1 - \frac{x}{L}\right)\left(2\frac{x}{L} - \frac{x^2}{L^2} - \frac{a^2}{L^2}\right) \quad \text{kip} \cdot \text{ft}^3$$

$$\Delta_x = \frac{PL^3}{6EI} \frac{a}{L}\left(1 - \frac{x}{L}\right)\left(2\frac{x}{L} - \frac{x^2}{L^2} - \frac{a^2}{L^2}\right) \quad (2\text{-}2)$$

Equations (2-1) and (2-2) are similar: by substituting a/L for x/L and vice versa in Eq. (2-1) it becomes Eq. (2-2).

A general equation may be written as

$$\Delta_x = \frac{PL^3}{EI} C \quad (2\text{-}3)$$

where
$$C = \begin{cases} \dfrac{1}{6}\dfrac{x}{L}\left(1 - \dfrac{a}{L}\right)\left(2\dfrac{a}{L} - \dfrac{a^2}{L^2} - \dfrac{x^2}{L^2}\right) & \text{when } x < a \\[2ex] \dfrac{1}{6}\dfrac{a}{L}\left(1 - \dfrac{x}{L}\right)\left(2\dfrac{x}{L} - \dfrac{x^2}{L^2} - \dfrac{a^2}{L^2}\right) & \text{when } x > a \end{cases}$$

When there is more than one concentrated load P, use $C_1, C_2, \ldots$ to represent each load condition and choose the formula according to whether the point of deflection x is smaller than or greater than a. The total deflection at a given point of the beam can be obtained by summing the deflection due to each load.

$$\Delta_x = \sum \frac{PL^3}{EI} C = \frac{P_1 L^3}{EI} C_1 + \frac{P_2 L^3}{EI} C_2 + \cdots \quad (2\text{-}4)$$

MAXIMUM DEFLECTION OF CONCENTRATED LOAD

For structural design, it is necessary to find the maximum deflection of a beam. Figure 2-2 shows the same beam as illustrated in Fig. 2-1. Let the maximum deflection occur at E, at a distance x from support A. The solution has two cases.

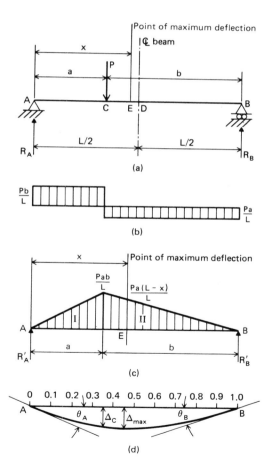

Fig. 2-2. (a) Given beam; (b) shear diagram; (c) moment diagram; (d) elastic curve.

Case 1: when $a < \dfrac{L}{2}$:

$$V'_E = 0 \qquad \frac{x}{L} = 1 - \sqrt{\frac{1}{3}\left(1 - \frac{a^2}{L^2}\right)} \tag{2-5}$$

$$M'_E = \frac{PL^3}{9}\frac{a}{L}\left(1 - \frac{a^2}{L^2}\right)\sqrt{\frac{1}{3}\left(1 - \frac{a^2}{L^2}\right)} \qquad \text{kip} \cdot \text{ft}^3$$

$$\Delta_{max} = \frac{PL^3}{9EI}\frac{a}{L}\left(1 - \frac{a^2}{L^2}\right)\sqrt{\frac{1}{3}\left(1 - \frac{a^2}{L^2}\right)} \tag{2-6}$$

Case 2: when $a > \dfrac{L}{2}$:

$$V'_E = 0 \qquad \frac{x}{L} = \sqrt{\frac{1}{3}\frac{a}{L}\left(2 - \frac{a}{L}\right)} \tag{2-7}$$

$$M'_E = \frac{PL}{9}\frac{a}{L}\left(1 - \frac{a}{L}\right)\left(2 - \frac{a}{L}\right)\sqrt{\frac{1}{3}\frac{a}{L}\left(2 - \frac{a}{L}\right)} \qquad \text{kip} \cdot \text{ft}^3$$

$$\Delta_{max} = \frac{PL^3}{9EI}\frac{a}{L}\left(1 - \frac{a}{L}\right)\left(2 - \frac{a}{L}\right)\sqrt{\frac{1}{3}\frac{a}{L}\left(2 - \frac{a}{L}\right)} \tag{2-8}$$

A general equation may be written as

$$\Delta_{max} = \frac{PL^3}{EI} C_{max} \tag{2-9}$$

where

$$C_{max} = \begin{cases} \dfrac{1}{9}\dfrac{a}{L}\left(1 - \dfrac{a^2}{L^2}\right)\sqrt{\dfrac{1}{3}\left(1 - \dfrac{a^2}{L^2}\right)} & \text{when } a < \dfrac{L}{2} \\[2ex] \dfrac{1}{9}\dfrac{a}{L}\left(1 - \dfrac{a}{L}\right)\left(2 - \dfrac{a}{L}\right)\sqrt{\dfrac{1}{3}\dfrac{a}{L}\left(2 - \dfrac{a}{L}\right)} & \text{when } a > \dfrac{L}{2} \end{cases}$$

The deflection at the point of load is

$$\Delta_C = \frac{Pa^2b^2}{3EIL} \tag{2-10}$$

When the load is at midspan, the deflection at the point of load is

$$\Delta_D = \frac{PL^3}{48EI} \tag{2-11}$$

This position produces the greatest deflection for a simple-support beam carrying a concentrated load on the span.

Table 2-1, at the end of the chapter, shows the coefficients of deflection of a concentrated load at each one-tenth point of the beam. The values are found by solving the limiting equations for C in Eq. (2-3). Deflection at the center varies symmetrically for loads on either side of the centerline of the beam. The area indicated by the boxes in the table is called the *maximum deflection region*. Each coefficient within the boxed areas is the largest coefficient for that point of load. Usually it occurs between $x/L = 0.4$ and $x/L = 0.6$.

Table 2-2, at the end of the chapter, lists the coefficient and position of maximum deflection for a simple-support beam carrying a concentrated load. The coefficients are computed by solving the limiting equations for C in Eq. (2-9). When there is more than one load, the position for the maximum deflection of the combined load has to be found. However, for design purposes, we can use point 0.4 or 0.5, whichever is closer to the maximum deflection, to calculate an approximate maximum deflection, rather than locate the exact position of the combined load to calculate the exact maximum deflection. The difference between the approximate and exact maximum deflections is very small. The largest error between the exact maximum deflection coefficient (Table 2-2) and the approximate maximum deflection coefficient (Table 2-1) occurs at $a/L = 0.3$ loading, and the percentage difference is $(0.016706 - 0.016500)/0.016706 = 1.23$ percent.

Example 2-1

A simple-support beam has a span of L ft. It carries a concentrated load of P at a distance $0.2L$ from support A. Find the maximum deflection and the approximate maximum deflection.

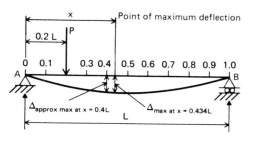

Fig. 2-3.

Solution

The maximum deflection occurs at $x/L = 0.434$ (from Table 2-2)

$$\Delta_{max} = \frac{PL^3}{EI} \times 0.012068 \text{ in downward} \qquad (ans.)$$

The approximate maximum deflection occurs at the point 0.4, and from Table 2-1

$$\Delta_{\text{approx. max}} = \frac{PL^3}{EI} \times 0.012000 \text{ in downward} \qquad (ans.)$$

The percentage error $= (0.12068 - 0.012000)/0.012068 = 0.56$ percent.

Example 2-2

A bridge stringer has a span of 82 ft. It carries a truck loading HS20-44 as shown in Fig. 2-4(a). Find the position and the value of the maximum deflection, and find the approximate maximum deflection. The stringer spacing is 7 ft 8 in, the load distribution factor is 0.697, and the impact of loads is 24 percent. The AASHTO standard specifications for highway bridges are used.

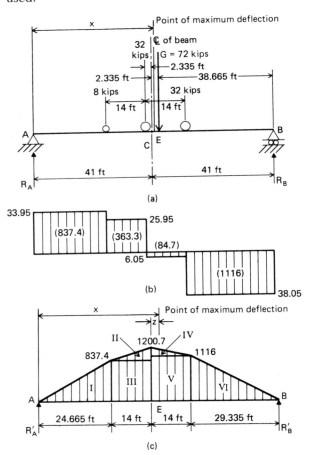

Fig. 2-4. (a) Given beam; (b) shear diagram; (c) moment diagram.

Solution

A. EXACT SOLUTION: Compute the exact position for the maximum deflection.

$$R_A = 72(41 - 2.335)/82 = 33.95 \text{ kips}$$
$$R_B = 72(41 + 2.335)/82 = 38.05 \text{ kips}$$

AREA OF MOMENT

$$\begin{aligned}
\text{Area I} &= \tfrac{1}{2} \times 837.4 \times 24.665 = 10{,}327 \text{ kip} \cdot \text{ft}^2 \\
\text{Area II} &= \tfrac{1}{2} \times 363.3 \times 14 = 2{,}543 \text{ kip} \cdot \text{ft}^2 \\
\text{Area III} &= 837.4 \times 14 = 11{,}724 \text{ kip} \cdot \text{ft}^2 \\
\text{Area IV} &= \tfrac{1}{2} \times 84.7 \times 14 = 593 \text{ kip} \cdot \text{ft}^2 \\
\text{Area V} &= 1{,}116 \times 14 = 15{,}624 \text{ kip} \cdot \text{ft}^2 \\
\text{Area VI} &= \tfrac{1}{2} \times 1{,}116 \times 29.335 = \underline{16{,}369 \text{ kip} \cdot \text{ft}^2} \\
& \hspace{4em} 57{,}180 \text{ kip} \cdot \text{ft}^2
\end{aligned}$$

$$R'_A = 28{,}048 \text{ kip} \cdot \text{ft}^2$$
$$R'_B = 29{,}132 \text{ kip} \cdot \text{ft}^2$$
$$V'_E = 0$$

$$R'_A - \text{Area I} - \text{Area II} - \text{Area III}$$

$$- 1{,}116z - 84.7\left(1 - \frac{z}{14}\right)z - \frac{1}{2} \times 84.7 \frac{z^2}{14} = 0$$

Solving for z

$$z = 2.898 \text{ ft} \quad \text{and} \quad x = 41.563 \text{ ft} \quad (ans.)$$

then

$$\begin{aligned}
M'_E = R'_A &\times 41.563 - \text{Area I} \times 25.120 \\
&- \text{Area II} \times 7.565 - \text{Area III} \times 9.898 \\
&- 1{,}200.7 \times \frac{2.898^2}{2} + 3.025 \times 2.898^2 \times \frac{1}{3} \times 2{,}898 \\
&= 766{,}046 \text{ kip} \cdot \text{ft}^3
\end{aligned}$$

$$\Delta_{\max}(\text{at point } E) = \frac{766{,}046 \times 1{,}728}{EI} \quad 0.697 \times 1.24$$

$$= \frac{1}{EI} \times 1.1441 \text{ in downward} \quad (ans.)$$

B. APPROXIMATE SOLUTION: Compute the maximum deflection at $x/L = 0.5$ using Table 2-1.

Load, kips	Distance from A, ft	Point of load	Value of C at $x/L = 0.5$
$P_1 = 8$	$a_1 = 24.665$	$\dfrac{a_1}{L} = 0.301$	0.016538
$P_2 = 32$	$a_2 = 38.665$	$\dfrac{a_2}{L} = 0.471$	0.020721
$P_3 = 32$	$a_3 = 52.665$	$\dfrac{a_3}{L} = 0.642$	0.018539

$$\text{Approximate maximum } \Delta_{0.5} = \frac{82^3 \times 1728}{EI}$$

$$\times (8 \times 0.016538 + 32 \times 0.020721 + 32 \times 0.018539)$$

$$= \frac{1}{EI} \times 1.1435 \text{ in downward} \quad (ans.)$$

The two results are almost equal because the position of the maximum deflection is close to the center of the span.

2-3 SIMPLE-SUPPORT BEAM WITH UNIFORMLY DISTRIBUTED LOAD

To find a general equation for this case, we assume that a simple-support beam AB carries a partial, uniformly distributed load on the span and that the load extends from the left support toward the right support as in Fig. 2-5. We wish to determine the deflection of the beam for each extended load.

From Fig. 2-5(b), we obtain

$$R_A = V_A = \frac{wa(2L - a)}{2L} \quad \text{kips}$$

$$R_B = V_B = \frac{wa^2}{2L} \quad \text{kips}$$

When $x < a$:

$$M_x = V_A x - \frac{1}{2} wx^2 = \frac{wa(2L - a)x}{2L} - \frac{1}{2} wx^2 \quad \text{kip} \cdot \text{ft}$$

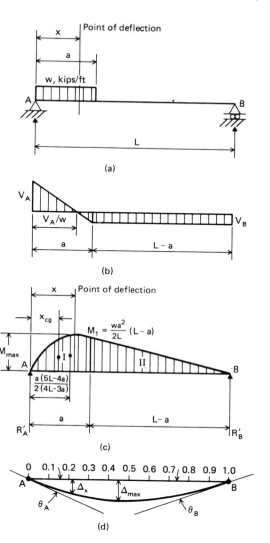

Fig. 2-5. (a) Given beam; (b) shear diagram; (c) moment diagram; (d) elastic curve.

When $x = a$:

$$M_1 = \frac{a}{2}(V_A - V_B) = \frac{wa^2}{2L}(L - a) \quad \text{kip} \cdot \text{ft}$$

The maximum moment occurs when $V = 0$ and $x = V_A/w$. Thus

$$M_{max} = \frac{V_A^2}{2w} = \frac{w}{8}\left[\frac{a(2L - a)}{L}\right]^2 \quad \text{kip} \cdot \text{ft}$$

24 Chapter 2

The center of gravity of the moment area for a partial, uniformly distributed load, x_{cg}, as in Fig. 2-5(c), can be found as follows:

$$\int_0^x M \, dx = \frac{wa(2L-a)x^2}{4L} - \frac{1}{6}wx^3 \quad \text{kip} \cdot \text{ft}^2 \tag{2-12}$$

$$\int_0^x Mx \, dx = \frac{wa(2L-a)x^3}{6L} - \frac{1}{8}wx^4 \quad \text{kip} \cdot \text{ft}^3 \tag{2-13}$$

$$x_{cg} = \frac{\int Mx \, dx}{\int M \, dx} = \frac{4a(2L-a)x - 3x^2 L}{6a(2L-a) - 4xL} \quad \text{ft} \tag{2-14}$$

When $x = a$:

$$x_{cg} = \frac{a(5L-4a)}{2(4L-3a)} \quad \text{ft} \tag{2-15}$$

Substituting a for x in Eq. (2-12), we obtain

$$\text{Area I} = \frac{wa^3}{12L}(4L-3a) \quad \text{kip} \cdot \text{ft}^2$$

and from M_1 and $(L-a)$, we obtain

$$\text{Area II} = \frac{wa^2}{4L}(L-a)^2 \quad \text{kip} \cdot \text{ft}^2$$

$$R'_A = \frac{wa^3}{24L^2}(8L^2 - 11aL + 4a^2) + \frac{wa^2}{6L^2}(L-a)^3 \quad \text{kip} \cdot \text{ft}^2$$

$$R'_B = \frac{wa^4}{24L^2}(5L-4a) + \frac{wa^2}{12L^2}(L-a)^2(2a+L) \quad \text{kip} \cdot \text{ft}^2$$

When $x < a$:

$$M'_x = R'_A x - \int M \, dx \, (x - x_{cg})$$

$$= \frac{wx}{24L}[a^2(2L-a)^2 - 2a(2L-a)x^2 + Lx^3] \quad \text{kip} \cdot \text{ft}^3$$

$$\Delta_x = \frac{wL^4}{24EI}\frac{x}{L}\left[\frac{a^2}{L^2}\left(2 - \frac{a}{L}\right)^2 - 2\frac{a}{L}\left(2 - \frac{a}{L}\right)\frac{x^2}{L^2} + \frac{x^3}{L^3}\right] \tag{2-16}$$

When $x > a$:

$$M'_x = R'_B(L-x) - \frac{1}{6}\frac{M_1(L-x)^3}{(L-a)}$$

$$= \frac{wa^2}{24}\left[\left(1 - \frac{x}{L}\right)(4Lx - 2x^2 - a^2)\right] \quad \text{kip} \cdot \text{ft}^3$$

$$\Delta_x = \frac{wL^4}{24EI}\frac{a^2}{L^2}\left[\left(1 - \frac{x}{L}\right)\left(4\frac{x}{L} - 2\frac{x^2}{L^2} - \frac{a^2}{L^2}\right)\right] \tag{2-17}$$

Deflection of Simple-Support Beams

A general equation may be written as

$$\Delta_x = \frac{wL^4}{EI} C \tag{2-18}$$

where

$$C = \begin{cases} \dfrac{1}{24}\dfrac{x}{L}\left[\dfrac{a^2}{L^2}\left(2 - \dfrac{a}{L}\right)^2 - 2\dfrac{a}{L}\left(2 - \dfrac{a}{L}\right)\dfrac{x^2}{L^2} + \dfrac{x^3}{L^3}\right] & \text{when } x < a \\[1em] \dfrac{1}{24}\dfrac{a^2}{L^2}\left[\left(1 - \dfrac{x}{L}\right)\left(4\dfrac{x}{L} - 2\dfrac{x^2}{L^2} - \dfrac{a^2}{L^2}\right)\right] & \text{when } x > a \end{cases}$$

When the uniformly distributed load is over the whole span, i.e., when $a = L$, then by Eq. (2-18), for the condition $x < a$,

$$\Delta_x = \frac{wL^4}{24EI}\frac{x}{L}\left(1 - 2\frac{x^2}{L^2} + \frac{x^3}{L^3}\right) \tag{2-19}$$

where

$$C = \frac{1}{24}\frac{x}{L}\left(1 - 2\frac{x^2}{L^2} + \frac{x^3}{L^3}\right)$$

The deflection at the center of the span can be found by setting $x/L = 0.5$ in Eq. (2-19)

$$\Delta_{0.5} = \frac{5wL^4}{384EI} \tag{2-20}$$

The coefficients of deflection for Eq. (2-18) are listed in Table 2-3 at the end of the chapter. The region of largest deflection occurs between the points 0.4 and 0.5. The maximum deflection will occur at the center when the uniformly distributed load is over the whole span or symmetrically distributed about the centerline of the beam.

Table 2-3 is very useful. We can find the deflection for any part of a uniformly distributed load over the span by subtracting the deflection of one partial, uniformly distributed load from the deflection of a larger partial, uniformly distributed load. The table assumes that the uniformly distributed load extends from the left support to the right support. When the uniformly distributed load extends from the right support to the left support, read the values of C in Table 2-3 at the opposite direction, that is, under the complement of the deflection point x/L under consideration. For example, a deflection at $x/L = 0.4$ is read at deflection $x/L = 0.6$. This is illustrated in Example 2-4.

MAXIMUM DEFLECTION OF UNIFORMLY DISTRIBUTED LOAD

The best solution for determining the position of the maximum deflection for a beam carrying a uniformly distributed load extending from one end toward the other is to differentiate the values of C and set $\partial C/\partial x = 0$ at each point of the loading. The solutions for C are listed at the end of the chapter in Table 2-4 and are based on the equation

$$\Delta_{max} = \frac{wL^4}{EI} C_{max} \qquad (2\text{-}21)$$

The maximum deflection of a simple-support beam with a uniformly distributed load may also be found by an approximate method. The biggest difference between the maximum deflection (from Table 2-4) and the approximate maximum deflection (from Table 2-3) occurs at $0.425L$ loading. The percentage error is $(0.005027 - 0.004965)/0.005027 = 1.23$ percent.

Example 2-3

Find the maximum deflection for a simple-support beam carrying a partial, uniformly distributed load at the center of the beam as shown in Fig. 2-6. Find Δ_{max} for the conditions $c = 0.2L$, $0.4L$, and $0.6L$. (Use Table 2-3.)

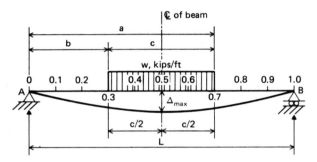

Fig. 2-6.

Solution

Since the load is distributed symmetrically about the centerline of the beam, the maximum deflection occurs at the center. The solution of this problem can be found by subtracting the partial, uniformly distributed load b from the partial, uniformly distributed load a to obtain the maximum deflection of a uniformly distributed load c at the center.

When $c = 0.2L$: $\quad \Delta_{max} = \dfrac{wL^4}{EI}(0.008554 - 0.004467)$

$\qquad\qquad\qquad\qquad = \dfrac{wL^4}{EI} 0.004087$ in downward $\qquad$ (ans.)

When $c = 0.4L$: $\quad \Delta_{max} = \dfrac{wL^4}{EI}(0.010377 - 0.002644)$

$\qquad\qquad\qquad\qquad = \dfrac{wL^4}{EI} 0.007733$ in downward $\qquad$ (ans.)

When $c = 0.6L$:
$$\Delta_{max} = \frac{wL^4}{EI}(0.011804 - 0.001217)$$
$$= \frac{wL^4}{EI} 0.010587 \text{ in downward} \qquad (ans.)$$

Example 2-4

Find the position of the approximate maximum deflection of a simple-support beam carrying a partial, uniformly distributed load at both ends, as shown in Fig. 2-7. (Use Table 2-3.)

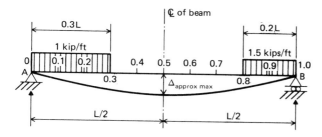

Fig. 2-7.

Solution

By inspection, the approximate maximum deflection occurs at point 0.4 or 0.5. The load of $1.5w$ kips/ft over a length $0.2L$ is on the right support. Read the values of C in the opposite direction in Table 2-3 instead of using the subtraction method illustrated in Example 2-3. For example, the value of C for a deflection at $x/L = 0.4$ is read at $x/L = 0.6$. The deflection of point 0.4 is

$$\Delta_{0.4} = \frac{wL^4}{EI} 0.002678 + \frac{1.5\, wL^4}{EI} 0.001093$$

$$= \frac{wL^4}{EI} 0.004317 \text{ in downward}$$

The deflection of point 0.5 is

$$\Delta_{0.5} = \frac{wL^4}{EI} 0.002644 + \frac{1.5\, wL^4}{EI} 0.001217$$

$$= \frac{wL^4}{EI} 0.004469 \text{ in downward}$$

The approximate maximum deflection is at point 0.5. $\qquad (ans.)$

TABLE 2-1 COEFFICIENTS OF DEFLECTION OF A SIMPLE-SUPPORT BEAM CARRYING A CONCENTRATED LOAD

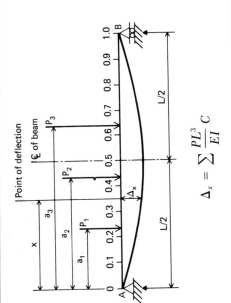

$$\Delta_x = \sum \frac{PL^3}{EI} C$$

Position of concentrated load, a/L	Coefficient C								
	$x/L = 0.1$	$x/L = 0.2$	$x/L = 0.3$	$x/L = 0.4$	$x/L = 0.5$	$x/L = 0.6$	$x/L = 0.7$	$x/L = 0.8$	$x/L = 0.9$
0.025	0.000710	0.001198	0.001486	0.001598	0.001561	0.001399	0.001137	0.000799	0.000412
0.050	0.001406	0.002383	0.002960	0.003187	0.003115	0.002792	0.002269	0.001596	0.000823
0.075	0.002074	0.003544	0.004413	0.004758	0.004652	0.004172	0.003391	0.002386	0.001230
0.100	0.002700	0.004667	0.005833	0.006300	0.006167	0.005533	0.004500	0.003167	0.001633
0.125	0.003272	0.005740	0.007210	0.007805	0.007650	0.006870	0.005590	0.003935	0.002030
0.150	0.003790	0.006750	0.008531	0.009262	0.009094	0.008175	0.006656	0.004687	0.002419
0.175	0.004254	0.007685	0.009787	0.010664	0.010491	0.009443	0.007695	0.005421	0.002798
0.200	0.004667	0.008533	0.010967	0.012000	0.011833	0.010667	0.008700	0.006133	0.003167
0.225	0.005029	0.009284	0.012059	0.013261	0.013113	0.011841	0.009668	0.006820	0.003523
0.250	0.005344	0.009937	0.013052	0.014438	0.014323	0.012958	0.010594	0.007479	0.003865
0.275	0.005611	0.010497	0.013936	0.015520	0.015454	0.014014	0.011473	0.008107	0.004191
0.300	0.005833	0.010967	0.014700	0.016500	0.016500	0.015000	0.012300	0.008700	0.004500

Position									
0.325	0.006012	0.011348	0.015335	0.017367	0.017452	0.015911	0.013071	0.009256	0.004790
0.350	0.006148	0.011646	0.015844	0.018112	0.018302	0.016742	0.013781	0.009771	0.005060
0.375	0.006243	0.011862	0.016230	0.018727	0.019043	0.017484	0.014426	0.010242	0.005309
0.400	0.006300	0.012000	0.016500	0.019200	0.019667	0.018133	0.015000	0.010667	0.005533
0.425	0.006319	0.012063	0.016657	0.019526	0.020165	0.018682	0.015499	0.011041	0.005733
0.450	0.006302	0.012054	0.016706	0.019708	0.020531	0.019125	0.015919	0.011363	0.005906
0.475	0.006251	0.011977	0.016652	0.019753	0.020757	0.019455	0.016254	0.011628	0.006051
0.500	0.006167	0.011833	0.016500	0.019667	0.020833	0.019667	0.016500	0.011833	0.006167
0.525	0.006051	0.011628	0.016254	0.019455	0.020757	0.019753	0.016652	0.011977	0.006251
0.550	0.005906	0.011363	0.015919	0.019125	0.020531	0.019708	0.016706	0.012054	0.006302
0.575	0.005733	0.011041	0.015499	0.018682	0.020165	0.019526	0.016657	0.012063	0.006319
0.600	0.005533	0.010667	0.015000	0.018133	0.019667	0.019200	0.016500	0.012000	0.006300
0.625	0.005309	0.010242	0.014426	0.017484	0.019043	0.018727	0.016230	0.011862	0.006243
0.650	0.005060	0.009771	0.013781	0.016742	0.018302	0.018112	0.015844	0.011646	0.006148
0.675	0.004790	0.009256	0.013071	0.015911	0.017452	0.017367	0.015335	0.011348	0.006012
0.700	0.004500	0.008700	0.012300	0.015000	0.016500	0.016500	0.014700	0.010967	0.005833
0.725	0.004191	0.008107	0.011473	0.014014	0.015454	0.015520	0.013936	0.010497	0.005611
0.750	0.003865	0.007479	0.010594	0.012958	0.014323	0.014438	0.013052	0.009937	0.005344
0.775	0.003523	0.006820	0.009668	0.011841	0.013113	0.013261	0.012059	0.009284	0.005029
0.800	0.003167	0.006133	0.008700	0.010667	0.011833	0.012000	0.010967	0.008533	0.004667
0.825	0.002798	0.005421	0.007695	0.009443	0.010491	0.010664	0.009787	0.007685	0.004254
0.850	0.002419	0.004687	0.006656	0.008175	0.009094	0.009262	0.008531	0.006750	0.003790
0.875	0.002030	0.003935	0.005590	0.006870	0.007650	0.007805	0.007210	0.005740	0.003272
0.900	0.001633	0.003167	0.004500	0.005533	0.006167	0.006300	0.005833	0.004667	0.002700
0.925	0.001230	0.002386	0.003391	0.004172	0.004652	0.004758	0.004413	0.003544	0.002074
0.950	0.000823	0.001596	0.002269	0.002792	0.003115	0.003188	0.002960	0.002383	0.001406
0.975	0.000412	0.000799	0.001137	0.001399	0.001561	0.001598	0.001486	0.001198	0.000710
1.000	0.000000	0.000000	0.000000	0.000000	0.000000	0.000000	0.000000	0.000000	0.000000

Note: The boxed areas contain the largest coefficient of deflection for each load position.

TABLE 2-2 MAXIMUM COEFFICIENTS OF DEFLECTION OF A SIMPLE-SUPPORT BEAM CARRYING A CONCENTRATED LOAD

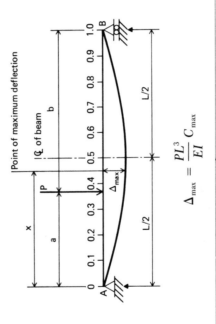

$$\Delta_{max} = \frac{PL^3}{EI} C_{max}$$

Position of concentrated load, a/L	Maximum coefficient, C_{max}	Position of max deflection, x/L	Position of concentrated load, a/L	Maximum coefficient, C_{max}	Position of max deflection, x/L
0.025	0.001602	0.423	0.525	0.020764	0.508
0.050	0.003195	0.423	0.550	0.020559	0.516
0.075	0.004771	0.424	0.575	0.020221	0.523
0.100	0.006319	0.426	0.600	0.019755	0.529

0.125	0.007832	0.427	
0.150	0.009300	0.429	
0.175	0.010715	0.432	
0.200	0.012068	0.434	
0.225	0.013352	0.437	
0.250	0.014558	0.441	
0.275	0.015678	0.445	
0.300	0.016706	0.449	
0.325	0.017634	0.454	
0.350	0.018456	0.459	
0.375	0.019165	0.465	
0.400	0.019755	0.471	
0.425	0.020221	0.477	
0.450	0.020559	0.484	
0.475	0.020764	0.492	
0.500	0.020833	0.500	
0.625	0.019165		0.535
0.650	0.018456		0.541
0.675	0.017634		0.546
0.700	0.016706		0.551
0.725	0.015678		0.555
0.750	0.014558		0.559
0.775	0.013352		0.563
0.800	0.012068		0.566
0.825	0.010715		0.568
0.850	0.009300		0.571
0.875	0.007832		0.573
0.900	0.006319		0.574
0.925	0.004771		0.576
0.950	0.003195		0.577
0.975	0.001602		0.577
1.000	0.000000		0.000

TABLE 2-3 COEFFICIENTS OF DEFLECTION OF A SIMPLE-SUPPORT BEAM CARRYING A UNIFORMLY DISTRIBUTED LOAD EXTENDING FROM ONE END

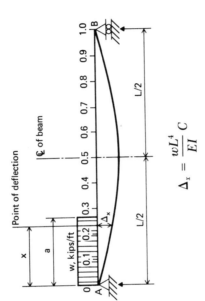

$$\Delta_x = \frac{wL^4}{EI} C$$

Length of uniform load extending from left support, a/L*	Coefficient C								
	$x/L = 0.1$	$x/L = 0.2$	$x/L = 0.3$	$x/L = 0.4$	$x/L = 0.5$	$x/L = 0.6$	$x/L = 0.7$	$x/L = 0.8$	$x/L = 0.9$
0.025	0.000009	0.000015	0.000019	0.000020	0.000020	0.000017	0.000014	0.000010	0.000005
0.050	0.000035	0.000060	0.000074	0.000080	0.000078	0.000070	0.000057	0.000040	0.000021
0.075	0.000079	0.000134	0.000166	0.000179	0.000175	0.000157	0.000128	0.000090	0.000046
0.100	0.000139	0.000237	0.000295	0.000318	0.000310	0.000278	0.000226	0.000159	0.000082
0.125	0.000214	0.000367	0.000458	0.000494	0.000483	0.000433	0.000352	0.000248	0.000128
0.150	0.000302	0.000523	0.000655	0.000707	0.000693	0.000622	0.000506	0.000356	0.000184
0.175	0.000403	0.000704	0.000884	0.000957	0.000937	0.000842	0.000685	0.000482	0.000249
0.200	0.000514	0.000907	0.001143	0.001240	0.001217	0.001093	0.000890	0.000627	0.000323
0.225	0.000635	0.001130	0.001431	0.001556	0.001529	0.001375	0.001120	0.000789	0.000407
0.250	0.000765	0.001370	0.001745	0.001902	0.001872	0.001685	0.001373	0.000967	0.000499
0.275	0.000902	0.001626	0.002083	0.002277	0.002244	0.002022	0.001649	0.001162	0.000600
0.300	0.001045	0.001894	0.002441	0.002678	0.002644	0.002385	0.001946	0.001373	0.000709

x/L									
0.325	0.001194	0.002173	0.002817	0.003101	0.003068	0.002772	0.002264	0.001597	0.000825
0.350	0.001346	0.002461	0.003207	0.003545	0.003515	0.003180	0.002599	0.001835	0.000948
0.375	0.001501	0.002755	0.003608	0.004006	0.003983	0.003608	0.002952	0.002085	0.001078
0.400	0.001658	0.003053	0.004018	0.004480	0.004467	0.004053	0.003320	0.002347	0.001213
0.425	0.001815	0.003354	0.004432	0.004964	0.004965	0.004514	0.003701	0.002618	0.001354
0.450	0.001973	0.003656	0.004849	0.005455	0.005474	0.004987	0.004094	0.002898	0.001500
0.475	0.002130	0.003956	0.005267	0.005949	0.005990	0.005469	0.004497	0.003186	0.001649
0.500	0.002285	0.004254	0.005681	0.006442	0.006510	0.005958	0.004906	0.003479	0.001802
0.525	0.002438	0.004548	0.006091	0.006931	0.007031	0.006451	0.005321	0.003777	0.001957
0.550	0.002588	0.004835	0.006493	0.007413	0.007547	0.006945	0.005738	0.004077	0.002114
0.575	0.002733	0.005115	0.006886	0.007886	0.008056	0.007436	0.006155	0.004379	0.002272
0.600	0.002874	0.005387	0.007267	0.008347	0.008554	0.007920	0.006570	0.004680	0.002430
0.625	0.003010	0.005648	0.007635	0.008792	0.009038	0.008394	0.006979	0.004978	0.002587
0.650	0.003139	0.005898	0.007988	0.009220	0.009505	0.008855	0.007381	0.005272	0.002742
0.675	0.003263	0.006136	0.008324	0.009628	0.009952	0.009299	0.007771	0.005560	0.002894
0.700	0.003379	0.006361	0.008641	0.010015	0.010377	0.009723	0.008146	0.005839	0.003042
0.725	0.003487	0.006571	0.008939	0.010378	0.010777	0.010123	0.008504	0.006108	0.003185
0.750	0.003588	0.006766	0.009214	0.010715	0.011149	0.010498	0.008842	0.006363	0.003322
0.775	0.003681	0.006945	0.009468	0.011025	0.011492	0.010844	0.009156	0.006604	0.003452
0.800	0.003764	0.007107	0.009697	0.011307	0.011804	0.011160	0.009444	0.006827	0.003573
0.825	0.003839	0.007251	0.009903	0.011558	0.012083	0.011443	0.009704	0.007030	0.003685
0.850	0.003904	0.007378	0.010082	0.011778	0.012328	0.011693	0.009933	0.007210	0.003786
0.875	0.003960	0.007485	0.010235	0.011967	0.012538	0.011906	0.010130	0.007366	0.003874
0.900	0.004005	0.007574	0.010361	0.012122	0.012710	0.012083	0.010293	0.007497	0.003949
0.925	0.004041	0.007644	0.010460	0.012243	0.012846	0.012221	0.010421	0.007599	0.004009
0.950	0.004067	0.007693	0.010531	0.012330	0.012943	0.012320	0.010513	0.007674	0.004052
0.975	0.004082	0.007723	0.010573	0.012382	0.013001	0.012380	0.010569	0.007718	0.004079
1.000	0.004088	0.007733	0.010587	0.012400	0.013021	0.012400	0.010587	0.007733	0.004088

*When the load extends from the right end, read C at the opposite (complementary) x/L point.

Note: The boxed areas contain the largest coefficient of deflection for each load position.

TABLE 2-4 MAXIMUM COEFICIENT OF DEFLECTION OF A SIMPLE-SUPPORT BEAM CARRYING A UNIFORMLY DISTRIBUTED LOAD EXTENDING FROM ONE END

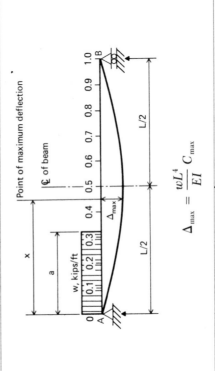

$$\Delta_{max} = \frac{wL^4}{EI} C_{max}$$

Length of uniform load extending from left support, a/L	Maximum coefficient, C_{max}	Position of max deflection, x/L	Length of uniform load extending from left support, a/L	Maximum coefficient, C_{max}	Position of max deflection, x/L
0.025	0.000020	0.423	0.525	0.007078	0.463
0.050	0.000080	0.423	0.550	0.007589	0.467
0.075	0.000180	0.423	0.575	0.008092	0.470
0.100	0.000318	0.424	0.600	0.008584	0.474

0.125	0.000495	0.425	0.625	0.009063	0.477
0.150	0.000710	0.426	0.650	0.009525	0.480
0.175	0.000960	0.427	0.675	0.009968	0.482
0.200	0.001245	0.428	0.700	0.010389	0.485
0.225	0.001563	0.430	0.725	0.010785	0.487
0.250	0.001912	0.432	0.750	0.011155	0.490
0.275	0.002289	0.434	0.775	0.011496	0.492
0.300	0.002694	0.436	0.800	0.011807	0.493
0.325	0.003123	0.438	0.825	0.012085	0.495
0.350	0.003574	0.441	0.850	0.012329	0.496
0.375	0.004043	0.443	0.875	0.012538	0.497
0.400	0.004529	0.446	0.900	0.012711	0.498
0.425	0.005027	0.449	0.925	0.012846	0.499
0.450	0.005534	0.453	0.950	0.012943	0.500
0.475	0.006048	0.456	0.975	0.013001	0.500
0.500	0.006563	0.460	1.000	0.013021	0.500

3
Deflection of Beams other than Simple-Support Beams

EFFECT OF END MOMENTS

When a beam is restrained or fixed at the support, it means that no angular rotation, or change of slope, of the centroidal axis of the beam can occur at that support. The elastic curve will be horizontal there.

At the fixed support, there is a fixed-end moment produced which counteracts the angular rotation that tends to occur at the end of the beam deflecting under the loading. The fixed-end moment has the effect of reducing the deflection of the beam. It produces an upward deflection against the downward deflection due to the loading. Therefore, in the design, it is desirable to minimize the deflection of the beam by fully restraining or fixing it at the support. Several fundamental formulas for fixed-end moments are illustrated here.

Consider a beam AB that has a constant cross section and that is fixed at both ends. When the beam is loaded with a concentrated load P, as in Fig. 3-1, the fixed-end moments M_{AB} and M_{BA} at the ends A and B are

$$M_{AB} = \frac{Pab^2}{L^2} \quad \text{kip} \cdot \text{ft} \tag{3-1}$$

$$M_{BA} = \frac{Pa^2b}{L^2} \quad \text{kip} \cdot \text{ft} \tag{3-2}$$

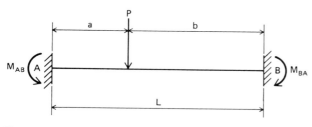

Fig. 3-1.

38 Chapter 3

When the beam has more than one concentrated load, the equations become

$$M_{AB} = \sum \frac{Pab^2}{L^2} \quad \text{kip} \cdot \text{ft} \tag{3-3}$$

$$M_{BA} = \sum \frac{Pa^2b}{L^2} \quad \text{kip} \cdot \text{ft} \tag{3-4}$$

in which a represents the left-side dimension and b represents the right-side dimension for each of the concentrated loads. When the beam is loaded with a uniformly distributed load over the whole span, as in Fig. 3-2, the fixed-end moments M_{AB} and M_{BA} at the ends A and B are equal.

$$M_{AB} = M_{BA} = \tfrac{1}{12}wL^2 \quad \text{kip} \cdot \text{ft} \tag{3-5}$$

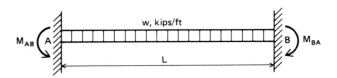

Fig. 3-2.

From these fundamental equations the equations of the fixed-end moments for any other loading condition can be derived.

Figure 3-3 shows a partial, uniformly distributed load, in which $P = w\,dx$, $a = x$, and $b = L - x$. The fixed-end moments M_{AB} and M_{BA} at the ends A and B are given by

$$M_{AB} = \int_0^{L/2} \frac{wx(L - x)^2\,dx}{L^2} = \frac{11\,wL^2}{192} \quad \text{kip} \cdot \text{ft} \tag{3-6}$$

$$M_{BA} = \int_0^{L/2} \frac{wx^2(L - x)\,dx}{L^2} = \frac{5\,wL^2}{192} \quad \text{kip} \cdot \text{ft} \tag{3-7}$$

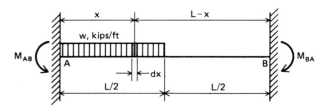

Fig. 3-3.

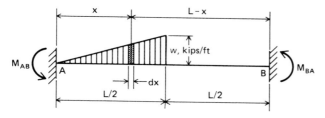

Fig. 3-4.

Figure 3-4 shows a varying load, increasing from A to the centerline of the beam, in which $P = (2wx\,dx)/L$, $a = x$, and $b = L - x$. The fixed-end moments M_{AB} and M_{BA} at the ends A and B are given by

$$M_{AB} = \int_0^{L/2} \frac{(2wx/L)x(L-x)^2\,dx}{L^2} = \frac{wL^2}{30} \quad \text{kip}\cdot\text{ft} \quad (3\text{-}8)$$

$$M_{BA} = \int_0^{L/2} \frac{(2wx/L)x^2(L-x)\,dx}{L^2} = \frac{3wL^2}{160} \quad \text{kip}\cdot\text{ft} \quad (3\text{-}9)$$

In the same manner, when the load decreases from A to the centerline of the beam, as shown in Fig. 3-5, where $P = [(L - 2x)/L]w\,dx$, $a = x$, and $b = L - x$, the fixed-end moments M_{AB} and M_{BA} at the ends A and B are given by

$$M_{AB} = \int_0^{L/2} \frac{[(L-2x)/L]x(L-x)^2 w\,dx}{L^2} = \frac{23wL^2}{960} \quad \text{kip}\cdot\text{ft} \quad (3\text{-}10)$$

$$M_{BA} = \int_0^{L/2} \frac{[(L-2x)/L]x^2(L-x)w\,dx}{L^2} = \frac{7wL^2}{960} \quad \text{kip}\cdot\text{ft} \quad (3\text{-}11)$$

In calculating the shear force to determine the positive moment for a beam fixed at both ends, a simple rule to remember is that the shear force

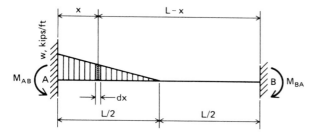

Fig. 3-5.

40 *Chapter 3*

is greater than the simple-support reaction at the fixed end having the greater numerical fixed-end moment.

The foregoing are examples of very common loadings. Table 3-1, at the end of the chapter, lists formulas for fixed-end moments under various loading conditions, including cases with hinges on either end. From these loadings the fixed-end moments of other loadings can be obtained by the superposition method, adding or subtracting to obtain the desired values of M_{AB} and M_{BA}.

3-2 DEFLECTION OF BEAMS DUE TO END MOMENTS

Assume that a beam has the end moments M_{AB} and M_{BA} as shown in Fig. 3-6(a) and its equivalent diagrams (b) and (c).

In (b) $\quad R'_A = \tfrac{1}{3} M_{AB} L \quad$ kip · ft^2

$\quad\quad\quad R'_B = \tfrac{1}{6} M_{AB} L \quad$ kip · ft^2

In (c) $\quad R'_A = \tfrac{1}{6} M_{BA} L \quad$ kip · ft^2

$\quad\quad\quad R'_B = \tfrac{1}{3} M_{BA} L \quad$ kip · ft^2

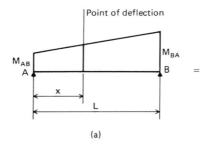

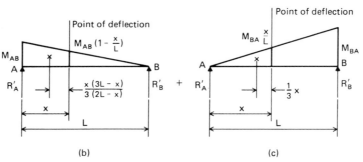

Fig. 3-6. (a) End moments diagram; (b) end moment M_{AB} diagram; (c) end moment M_{BA} diagram.

Deflection of Beams other than Simple-Support Beams 41

These moment diagrams represent the load of the moment area on a conjugate beam. The moment at any point x on the conjugate beam is the deflection at that point in the actual beam. The deflection due to the end moment M_{AB} at point x, Fig. 3-6(b), is

$$M'_x = -\frac{M_{AB}L^2}{6}\frac{x}{L}\left(2 - 3\frac{x}{L} + \frac{x^2}{L^2}\right) \quad \text{kip} \cdot \text{ft}^3$$

$$\Delta_x = -\frac{M_{AB}L^2}{6EI}\frac{x}{L}\left(2 - 3\frac{x}{L} + \frac{x^2}{L^2}\right)$$

and the deflection due to the end moment M_{BA} at the same point, Fig. 3-6(c), is

$$M'_x = -\frac{M_{BA}L^2}{6}\frac{x}{L}\left(1 - \frac{x^2}{L^2}\right) \quad \text{kip} \cdot \text{ft}^3$$

$$\Delta_x = -\frac{M_{BA}L^2}{6EI}\frac{x}{L}\left(1 - \frac{x^2}{L^2}\right)$$

A general equation for the deflection of a beam due to the end moments at both ends is

$$\Delta_x = -\frac{L^2}{EI}(M_{AB}C_{AB} + M_{BA}C_{BA}) \tag{3-12}$$

where

$$C_{AB} = \frac{1}{6}\frac{x}{L}\left(2 - 3\frac{x}{L} + \frac{x^2}{L^2}\right)$$

$$C_{BA} = \frac{1}{6}\frac{x}{L}\left(1 - \frac{x^2}{L^2}\right)$$

The minus sign means the deflection is upward and is counteracting the downward deflection due to loadings. The coefficients of deflection at each one-tenth point of the beam due to the end moments are listed at the end of the chapter in Table 3-2.

The value of C_{AB} can be calculated by substituting $1 - x/L$ for x/L in the equation for calculating C_{BA}. When the end moments M_{AB} and M_{BA} are equal, the coefficients C_{AB} and C_{BA} may be added and become $C_{AB} + C_{BA}$. When either end moment M_{AB} or M_{BA} equals zero, the equation becomes the deflection of a beam, fixed at one end and supported at the other. When both end moments M_{AB} and M_{BA} are equal to zero, the beam becomes a simple-support beam.

3-3 DEFLECTION OF A BEAM, FIXED AT ONE END AND SUPPORTED AT THE OTHER

In this case we have three unknown elements at the fixed end and one unknown element at the supported end. This is a statically indeterminate beam to the first degree. By subtracting the deflection due to the end moments from the deflection due to the load on a simple-support beam, we can find the deflection of the beam.

Concentrated Load

In order to find a general equation for this case, we assume that the beam, fixed at one end and supported at the other, carries a concentrated load P as shown in Fig. 3-7(a). We wish to determine the deflection of the beam at any given point.

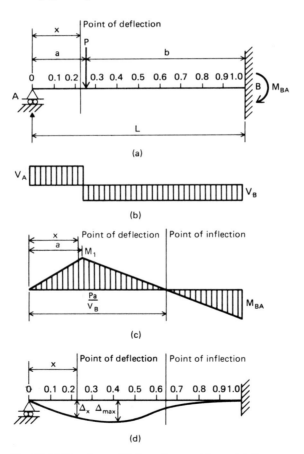

Fig. 3-7. (a) Given beam; (b) shear diagram; (c) moment diagram; (d) elastic curve.

Deflection of Beams other than Simple-Support Beams **43**

Because the beam is fixed at one end, there is a fixed-end moment produced there. It creates an upward deflection which counteracts the downward deflection due to loading. The point of inflection will be found at a point where the moment is zero. The curvature of deflection changes sign at that point.

The fixed-end moment for a beam with the left end hinged, loaded with a concentrated load, by Table 3-1, is

$$M_{BA} = -\frac{Pab}{2L^2}(L+a) \quad \text{kip} \cdot \text{ft}$$

$$R_A = V_A = \frac{P(L-a)^2}{2L^3}(2L+a) \quad \text{kips}$$

$$R_B = V_B = \frac{Pa}{2L^3}(3L^2 - a^2) \quad \text{kips}$$

$$M_1 = \frac{Pa(L-a)^2}{2L^3}(2L+a) \quad \text{kip} \cdot \text{ft}$$

The deflection of the beam due to the end moment M_{BA} at a distance x from support A, by Eq. (3-12), is

$$\Delta_x = -\frac{PL^3}{12EI}\frac{a}{L}\left(1-\frac{a^2}{L^2}\right)\frac{x}{L}\left(1-\frac{x^2}{L^2}\right) \quad (3\text{-}13)$$

The minus sign means the deflection is upward. By subtracting Eq. (3-13) from Eqs. (2-1) and (2-2) respectively, we obtain the final equations of the deflection of the beam.

$$\Delta_x = \begin{cases} \dfrac{PL^3}{12EI}\dfrac{x}{L}\left(1-\dfrac{a}{L}\right)^2\left[3\dfrac{a}{L}-\left(2+\dfrac{a}{L}\right)\dfrac{x^2}{L^2}\right] & \text{when } x < a \quad (3\text{-}14) \\ \dfrac{PL^3}{12EI}\dfrac{a}{L}\left(1-\dfrac{x}{L}\right)^2\left[\left(3-\dfrac{a^2}{L^2}\right)\dfrac{x}{L}-2\dfrac{a^2}{L^2}\right] & \text{when } x > a \quad (3\text{-}15) \end{cases}$$

A general equation may be written in the same form as that of Eq. (2-4)

$$\Delta_x = \sum \frac{PL^3}{EI} C \quad (3\text{-}16)$$

where

$$C = \begin{cases} \dfrac{1}{12}\dfrac{x}{L}\left(1-\dfrac{a}{L}\right)^2\left[3\dfrac{a}{L}-\left(2+\dfrac{a}{L}\right)\dfrac{x^2}{L^2}\right] & \text{when } x < a \\ \dfrac{1}{12}\dfrac{a}{L}\left(1-\dfrac{x}{L}\right)^2\left[\left(3-\dfrac{a^2}{L^2}\right)\dfrac{x}{L}-2\dfrac{a^2}{L^2}\right] & \text{when } x > a \end{cases}$$

The values of C are found by solving the limiting equations for C in Eq. (3-16) and are listed at the end of the chapter in Table 3-3. The maximum deflection of the beam can be found by differentiating the values of C and setting $\partial C/\partial x = 0$. The maximum coefficients of deflection for each loading are shown in Table 3-4 at the end of the chapter.

$$\Delta_{max} = \frac{PL^3}{EI} C_{max} \qquad (3\text{-}17)$$

where

$$C_{max} = \begin{cases} \dfrac{PL^3}{3EI} \dfrac{a}{L} \dfrac{(1 - a^2/L^2)^3}{(3 - a^2/L^2)^2} & \text{when } a < 0.414L \text{ at } x = L\dfrac{L^2 + a^2}{3L^2 - a^2} \\[2ex] \dfrac{PL^3}{6EI} \dfrac{a}{L} \left(1 - \dfrac{a}{L}\right)^2 \sqrt{\dfrac{a/L}{2 + a/L}} & \text{when } a > 0.414L \\[2ex] & \text{at } x = L\sqrt{\dfrac{a}{2L + a}} \end{cases}$$

Example 3-1

A single-span beam, fixed at one end and supported at the other, carries a concentrated load P at the center of the beam as shown in Fig. 3-8. Find the maximum deflection of the beam. (Use Table 3-4.)

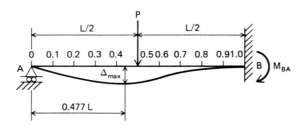

Fig. 3-8.

Solution

The maximum deflection is at $x/L = 0.447$ (from Table 3-4).

$$\Delta_{max} = \frac{PL^3}{EI} 0.009317 \text{ in downward} \qquad (ans.)$$

Example 3-2

A single-span beam, fixed at one end and supported at the other, carries a series of concentrated loads P_1, P_2, and P_3 on the span as shown in Fig. 3-9. Find the approximate maximum deflection of the beam. (Use Table 3-3.)

Deflection of Beams other than Simple-Support Beams

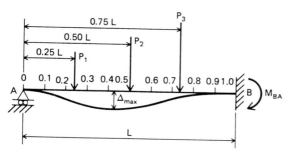

Fig. 3-9.

Solution

The maximum deflection occurs either at point 0.4 or at point 0.5, depending on the numerical values of the loads. The deflection at point 0.4 is

$$\Delta_{0.4} = \frac{L^3}{EI}(0.007875P_1 + 0.009167P_2 + 0.003771P_3)$$

The deflection at point 0.5 is

$$\Delta_{0.5} = \frac{L^3}{EI}(0.006999P_1 + 0.009115P_2 + 0.004069P_3)$$

When $P_1 = P_2 = P_3$, it may be seen that the deflection at point 0.4 is greater.

$$\Delta_{0.4} = \frac{PL^3}{EI} 0.020813 \text{ in downward} \qquad (ans.)$$

Uniformly Distributed Load

In order to find a general equation for this case, we assume that the beam, fixed at one end and supported at the other, carries a partial, uniformly distributed load, extending from the left support toward the right support, as shown in Fig. 3-10. We wish to determine the deflection of the beam for each extended load.

The fixed-end moment for a left-hinged beam loaded with a partial, uniformly distributed load, by Table 3-1, is

$$M_{BA} = -\frac{wa^2}{8}\left(2 - \frac{a^2}{L^2}\right) \quad \text{kip} \cdot \text{ft}$$

$$R_A = V_A = \frac{wa}{8}\left(8 - 6\frac{a}{L} + \frac{a^3}{L^3}\right) \quad \text{kips}$$

$$R_B = V_B = \frac{wa}{8}\left(6\frac{a}{L} - \frac{a^3}{L^3}\right) \quad \text{kips}$$

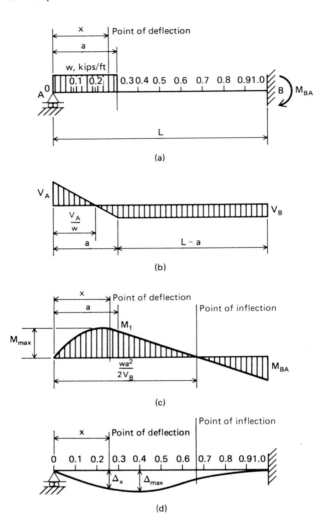

Fig. 3-10. (a) Given beam; (b) shear diagram; (c) moment diagram; (d) elastic curve.

When $x = a$:

$$M_1 = \frac{a}{2}(V_A - V_B) = \frac{wa^2}{8}\left(4 - 6\frac{a}{L} + \frac{a^3}{L^3}\right) \quad \text{kip} \cdot \text{ft}$$

The maximum deflection is at $x = V_A/w$

$$M_{max} = \frac{V_A^2}{2w} = \frac{wa^2}{128}\left(8 - 6\frac{a}{L} + \frac{a^3}{L^3}\right)^2 \quad \text{kip} \cdot \text{ft}$$

The deflection of the beam due to the fixed-end moment from the partial, uniformly distributed load can be found by substituting M_{BA} in Eq. (3-12)

$$\Delta_x = -\frac{wL^4}{48EI}\frac{a^2}{L^2}\frac{x}{L}\left(2-\frac{a^2}{L^2}\right)\left(1-\frac{x^2}{L^2}\right) \tag{3-18}$$

The minus sign means that the deflection is upward. Subtracting Eq. (3-18) from Eqs. (2-16) and (2-17) respectively, the final equations of the deflection of the beam are obtained.

$$\Delta_x = \begin{cases} \dfrac{wL^4}{24EI}\dfrac{x}{L}\left[\dfrac{a^2}{L^2}\left(2-\dfrac{a}{L}\right)^2 - 2\dfrac{a}{L}\left(2-\dfrac{a}{L}\right)\dfrac{x^2}{L^2} + \dfrac{x^3}{L^3} \right. \\ \left. \quad -\dfrac{1}{2}\dfrac{a^2}{L^2}\left(2-\dfrac{a^2}{L^2}\right)\left(1-\dfrac{x^2}{L^2}\right)\right] \quad \text{when } x < a \quad (3\text{-}19) \\[2ex] \dfrac{wL^4}{24EI}\dfrac{a^2}{L^2}\left[\left(1-\dfrac{x}{L}\right)\left(4\dfrac{x}{L}-2\dfrac{x^2}{L^2}-\dfrac{a^2}{L^2}\right) \right. \\ \left. \quad -\dfrac{1}{2}\dfrac{x}{L}\left(2-\dfrac{a^2}{L^2}\right)\left(1-\dfrac{x^2}{L^2}\right)\right] \quad \text{when } x > a \quad (3\text{-}20) \end{cases}$$

A general equation may be written in the same form as that of Eq. (2-18).

$$\Delta_x = \frac{wL^4}{EI}C \tag{3-21}$$

where

$$C = \begin{cases} \dfrac{1}{24}\dfrac{x}{L}\left[\dfrac{a^2}{L^2}\left(2-\dfrac{a}{L}\right)^2 - 2\dfrac{a}{L}\left(2-\dfrac{a}{L}\right)\dfrac{x^2}{L^2} + \dfrac{x^3}{L^3} \right. \\ \left. \quad -\dfrac{1}{2}\dfrac{a^2}{L^2}\left(2-\dfrac{a^2}{L^2}\right)\left(1-\dfrac{x^2}{L^2}\right)\right] \quad \text{when } x < a \\[2ex] \dfrac{1}{24}\dfrac{a^2}{L^2}\left[\left(1-\dfrac{x}{L}\right)\left(4\dfrac{x}{L}-2\dfrac{x^2}{L^2}-\dfrac{a^2}{L^2}\right) \right. \\ \left. \quad -\dfrac{1}{2}\dfrac{x}{L}\left(2-\dfrac{a^2}{L^2}\right)\left(1-\dfrac{x^2}{L^2}\right)\right] \quad \text{when } x > a \end{cases}$$

At the end of the chapter, Table 3-5 lists the coefficients C for a beam fixed at one end, supported at the other, and carrying a uniformly distributed load. In the above equations, the loading condition is the same as in the simple-support beam. We can use the same method to find the deflection for any part of a uniformly distributed load over the span by subtracting the deflection of a partial, uniformly distributed load from the deflection of a larger partial, uniformly distributed load as illustrated in Example 2-3.

In the same manner, the maximum coefficients of deflection for each extended load, listed at the end of the chapter in Table 3-6, can be found by

$$\Delta_{max} = \frac{wL^4}{EI} C_{max} \qquad (3\text{-}22)$$

When the uniformly distributed load is over the whole span, i.e., $a = L$, the deflection of the beam at any point, by Eq. (3-20), is

$$\Delta_x = \frac{wL^4}{48EI} \frac{x}{L}\left(1 - 3\frac{x^2}{L^2} + 2\frac{x^3}{L^3}\right) \qquad (3\text{-}23)$$

The deflection at the center of the beam is

$$\Delta_{0.5} = \frac{wL^4}{192EI} \qquad (3\text{-}24)$$

The maximum deflection of the beam is

$$\Delta_{max} = \frac{wL^4}{185EI} \quad \text{at } x = \frac{L}{16}(1 + \sqrt{33}) = 0.422L \qquad (3\text{-}25)$$

Example 3-3

Find the position of the approximate maximum deflection of a beam (see Fig. 3-11), fixed at one end and supported at the other, carrying a partial, uniformly distributed load at the center of the beam. Let $c = 0.2L, 0.4L,$ and $0.6L$. (Use Table 3-5.)

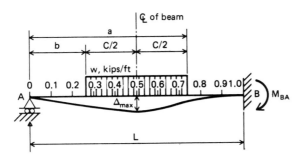

Fig. 3-11.

Solution

This problem may be solved by the same method as in Example 2-3, subtracting the partial, uniformly distributed load b from the partial, uniformly distributed load a to obtain the deflection of the partial, uniformly distributed load c.

When $c = 0.2L$:

$$\Delta_{0.4} = \frac{wL^4}{EI}(0.004214 - 0.002419) = \frac{wL^4}{EI} 0.001795 \text{ in downward}$$

$$\Delta_{0.5} = \frac{wL^4}{EI}(0.003942 - 0.002167) = \frac{wL^4}{EI} 0.001775 \text{ in downward}$$

When $c = 0.4L$:

$$\Delta_{0.4} = \frac{wL^4}{EI}(0.004836 - 0.001474) = \frac{wL^4}{EI} 0.003362 \text{ in downward}$$

$$\Delta_{0.5} = \frac{wL^4}{EI}(0.004597 - 0.001301) = \frac{wL^4}{EI} 0.003296 \text{ in downward}$$

When $c = 0.6L$:

$$\Delta_{0.4} = \frac{wL^4}{EI}(0.005214 - 0.000691) = \frac{wL^4}{EI} 0.004523 \text{ in downward}$$

$$\Delta_{0.5} = \frac{wL^4}{EI}(0.005004 - 0.000604) = \frac{wL^4}{EI} 0.004400 \text{ in downward}$$

The approximate maximum deflection occurs closer to the point 0.4 in each case of this beam. (ans.)

Example 3-4

Find the position of the approximate maximum deflection of a beam (see Fig. 3-12), fixed at one end and supported at the other, carrying a partial, uniformly distributed load $c = 0.2L$ at the fixed end. (Use Table 3-5.)

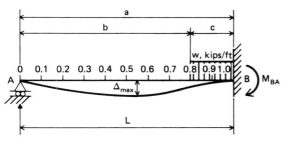

Fig. 3-12.

Solution

Try points 0.5 and 0.6 to find the approximate maximum deflection.

$$\Delta_{0.5} = \frac{wL^4}{EI}(0.005208 - 0.005004) = \frac{wL^4}{EI} 0.000204 \text{ in downward}$$

$$\Delta_{0.6} = \frac{wL^4}{EI}(0.004400 - 0.004197) = \frac{wL^4}{EI} 0.000203 \text{ in downward}$$

The approximate maximum deflection is at point 0.5. (ans.)

3-4 DEFLECTION OF A BEAM, FIXED AT BOTH ENDS

In this case we have three unknown elements at each fixed end. This is a statically indeterminate beam to the third degree. The solution of the deflection of the beam is as follows.

Concentrated Load

In order to derive a general equation for this case, we assume that a beam, fixed at both ends, carries a concentrated load P, as shown in Fig. 3-13. We wish to determine the deflection of the beam at any given point. By Table 3-1, we find the fixed-end moments

$$M_{AB} = \frac{Pa(L-a)^2}{L^2} \quad \text{kip} \cdot \text{ft}$$

$$M_{BA} = \frac{Pa^2(L-a)}{L^2} \quad \text{kip} \cdot \text{ft}$$

Substituting M_{AB} and M_{BA} in Eq. (3-12), we obtain the deflection of the beam due to the end moments for the concentrated load

$$\Delta_x = \frac{PL^3}{6EI}\frac{a}{L}\left(1 - \frac{a}{L}\right)$$

$$\times \frac{x}{L}\left[\left(1 - \frac{a}{L}\right)\left(2 - 3\frac{x}{L} + \frac{x^2}{L^2}\right) + \frac{a}{L}\left(1 - \frac{x^2}{L^2}\right)\right] \quad (3\text{-}26)$$

Subtracting Eq. (3-26) from Eqs. (2-1) and (2-2), we obtain the final equations of the deflection of the beam

$$\Delta_x = \begin{cases} \dfrac{PL^3}{6EI}\left(1 - \dfrac{a}{L}\right)^2 \dfrac{x^2}{L^2}\left[3\dfrac{a}{L} - \left(1 + 2\dfrac{a}{L}\right)\dfrac{x}{L}\right] & \text{when } x < a \quad (3\text{-}27) \\[2ex] \dfrac{PL^3}{6EI}\dfrac{a^2}{L^2}\dfrac{x^2}{L^2}\left[3\left(1 - \dfrac{a}{L}\right) - \left(3 - 2\dfrac{a}{L}\right)\dfrac{x}{L}\right] & \text{when } x > a \quad (3\text{-}28) \end{cases}$$

Deflection of Beams other than Simple-Support Beams 51

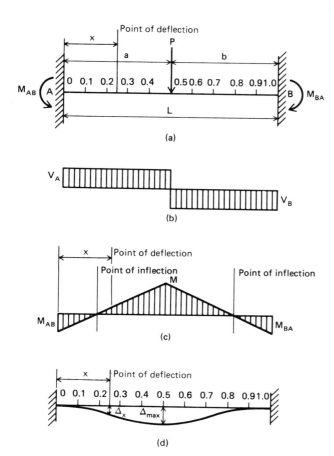

Fig. 3-13. (a) Given beam; (b) shear diagram; (c) moment diagram; (d) elastic curve.

A general equation may be written in the same form as that of Eq. (2-4)

$$\Delta_x = \sum \frac{PL^3}{EI} C \qquad (3\text{-}29)$$

where

$$C = \begin{cases} \dfrac{1}{6}\left(1 - \dfrac{a}{L}\right)^2 \dfrac{x^2}{L^2}\left[3\dfrac{a}{L} - \left(1 + 2\dfrac{a}{L}\right)\dfrac{x}{L}\right] & \text{when } x < a \\[2ex] \dfrac{1}{6}\dfrac{a^2}{L^2}\dfrac{x^2}{L^2}\left[3\left(1 - \dfrac{a}{L}\right) - \left(3 - 2\dfrac{a}{L}\right)\dfrac{x}{L}\right] & \text{when } x > a \end{cases}$$

The maximum deflection of the beam is given by

$$\Delta_{max} = \frac{PL^3}{EI} C_{max} \tag{3-30}$$

where

$$C_{max} = \begin{cases} \dfrac{2}{3}\dfrac{a^2}{L^2}\dfrac{(1 - a/L)^3}{(3 - 2a/L)^2} & \text{when } a < \dfrac{L}{2} \text{ at } x = \dfrac{L^2}{3L - 2a} \\ \dfrac{2}{3}\dfrac{a^3}{L^3}\dfrac{(1 - a/L)^2}{(1 + 2a/L)^2} & \text{when } a > \dfrac{L}{2} \text{ at } x = \dfrac{2aL}{L + 2a} \end{cases}$$

At the end of the chapter, Table 3-7 lists the coefficients C for a beam fixed at both ends and carrying a concentrated load, and Table 3-8 lists the maximum coefficients C_{max} for this situation. The maximum deflection occurs at the center when the load is placed at the center, i.e., $a = L/2$. By substituting this value in Eq. (3-30) we obtain

$$\Delta_{max} = \frac{PL^3}{192EI} \tag{3-31}$$

Example 3-5

A beam, fixed at both ends, carries a series of concentrated loads P_1, P_2, and P_3 on the span, as shown in Fig. 3-14. Find the approximate maximum deflection of the beam. (Use Table 3-7.)

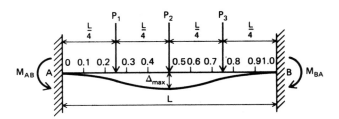

Fig. 3-14.

Solution

Try the deflection at points 0.4 and 0.5 and compare:

$$\Delta_{0.4} = \frac{L^3}{EI}(0.002813P_1 + 0.004667P_2 + 0.002083P_3)$$

$$\Delta_{0.5} = \frac{L^3}{EI}(0.002604P_1 + 0.005208P_2 + 0.002604P_3)$$

The approximate maximum deflection can occur at either point, depending on the actual loading.
When $P = P_1 = P_2 = P_3$, the maximum deflection is at point 0.5.

$$\Delta_{0.5} = \frac{PL^3}{EI} 0.010416 \text{ in downward} \qquad (ans.)$$

Example 3-6

A beam, fixed at both ends, carries a series of concentrated loads P_1, P_2, P_3, and P_4 on the span, as shown in Fig. 3-15. Find the approximate maximum deflection of the beam. (Use Table 3-7.)

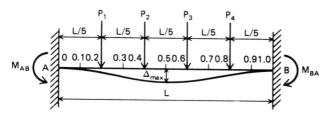

Fig. 3-15.

Solution

Try the deflection at points 0.4 and 0.5.

$$\Delta_{0.4} = \frac{L^3}{EI}(0.002016P_1 + 0.004608P_2 + 0.003925P_3 + 0.001451P_4)$$

$$\Delta_{0.5} = \frac{L^3}{EI}(0.001833P_1 + 0.004667P_2 + 0.004667P_3 + 0.001833P_4)$$

The approximate maximum deflection can occur at either point, depending on the actual loading. When $P_1 = P_2 = P_3 = P_4$, the maximum deflection is at point 0.5.

$$\Delta_{0.5} = \frac{PL^3}{EI} 0.01300 \text{ in downward} \qquad (ans.)$$

Uniformly Distributed Load

In order to derive a general equation for this case, we assume that a beam, fixed at both ends, carries a partial, uniformly distributed load extending from the left support, as shown in Fig. 3-16. We wish to determine the deflection of the beam for each extended load.

54 Chapter 3

From Table 3-1 we find the fixed-end moments

$$M_{AB} = \frac{wL^2}{12} \frac{a^2}{L^2}\left(6 - 8\frac{a}{L} + 3\frac{a^2}{L^2}\right) \quad \text{kip} \cdot \text{ft}$$

$$M_{BA} = \frac{wL^2}{12} \frac{a^3}{L^3}\left(4 - 3\frac{a}{L}\right) \quad \text{kip} \cdot \text{ft}$$

Substituting M_{AB} and M_{BA} in Eq. (3-12), we obtain the deflection of the beam due to the end moments from a partial, uniformly distributed load.

$$\Delta_x = -\frac{wL^4}{72EI} \frac{a^2}{L^2} \frac{x}{L}\left[\left(6 - 8\frac{a}{L} + 3\frac{a^2}{L^2}\right)\left(2 - 3\frac{x}{L} + \frac{x^2}{L^2}\right)\right.$$
$$\left. + \frac{a}{L}\left(4 - 3\frac{a}{L}\right)\left(1 - \frac{x^2}{L^2}\right)\right] \quad (3\text{-}32)$$

Subtracting Eq. (3-32) from Eqs. (2-16) and (2-17), respectively, we obtain the final equations of the deflection of the beam.

$$\Delta_x = \begin{cases} \dfrac{wL^4}{24EI} \dfrac{x}{L}\left[\dfrac{a^2}{L^2}\left(6 - 8\dfrac{a}{L} + 3\dfrac{a^2}{L^2}\right)\dfrac{x}{L} - \dfrac{a}{L}\left(4 - 4\dfrac{a^2}{L^2} + 2\dfrac{a^3}{L^3}\right)\dfrac{x^2}{L^2} + \dfrac{x^3}{L^3}\right] \\ \qquad\qquad\qquad\qquad\qquad\qquad\qquad\qquad \text{when } x < a \quad (3\text{-}33) \\[1em] \dfrac{wL^4}{24EI} \dfrac{a^2}{L^2}\left[-\dfrac{a^2}{L^2} + 4\dfrac{a}{L}\dfrac{x}{L} - \dfrac{a}{L}\left(8 - 3\dfrac{a}{L}\right)\dfrac{x^2}{L^2} + \dfrac{a}{L}\left(4 - 2\dfrac{a}{L}\right)\dfrac{x^3}{L^3}\right] \\ \qquad\qquad\qquad\qquad\qquad\qquad\qquad\qquad \text{when } x > a \quad (3\text{-}34) \end{cases}$$

A general equation may be written in the same form as Eq. (2-18)

$$\Delta_x = \frac{wL^4}{EI} C \qquad (3\text{-}35)$$

where

$$C = \begin{cases} \dfrac{1}{24} \dfrac{x}{L}\left[\dfrac{a^2}{L^2}\left(6 - 8\dfrac{a}{L} + 3\dfrac{a^2}{L^2}\right)\dfrac{x}{L} - \dfrac{a}{L}\left(4 - 4\dfrac{a^2}{L^2} + 2\dfrac{a^3}{L^3}\right)\dfrac{x^2}{L^2} + \dfrac{x^3}{L^3}\right] \\ \qquad\qquad\qquad\qquad\qquad\qquad\qquad\qquad \text{when } x < a \\[1em] \dfrac{1}{24} \dfrac{a^2}{L^2}\left[-\dfrac{a^2}{L^2} + 4\dfrac{a}{L}\dfrac{x}{L} - \dfrac{a}{L}\left(8 - 3\dfrac{a}{L}\right)\dfrac{x^2}{L^2} + \dfrac{a}{L}\left(4 - 2\dfrac{a}{L}\right)\dfrac{x^3}{L^3}\right] \\ \qquad\qquad\qquad\qquad\qquad\qquad\qquad\qquad \text{when } x > a \end{cases}$$

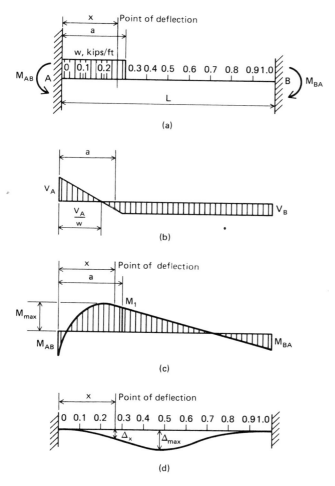

Fig. 3-16. (a) Given beam; (b) shear diagram; (c) moment diagram; (d) elastic curve.

The values of the coefficient C, for a beam fixed at both ends and carrying a uniform load extending from one end, are listed in Table 3-9 at the end of the chapter. The maximum deflection can be written in the same form as that of Eq. (3-22).

$$\Delta_{max} = \frac{wL^4}{EI} C_{max} \tag{3-36}$$

The maximum coefficients of deflection for each extended load are listed in Table 3-10 at the end of the chapter.

56 Chapter 3

When the uniformly distributed load is over the whole span, i.e., $a = L$, the deflection of the beam is given by

$$\Delta_x = \frac{wL^4}{24EI} \frac{x^2}{L^2} \left(1 - \frac{x}{L}\right)^2 \qquad (3\text{-}37)$$

The maximum deflection occurs at the center of the beam and is given by

$$\Delta_{max} = \frac{wL^4}{384EI} \qquad (3\text{-}38)$$

Example 3-7

Find the approximate maximum deflection of a beam, fixed at both ends, carrying a partial, uniformly distributed load at the center of the beam. Let $C = 0.2L, 0.4L$ and $0.6L$, as shown in Fig. 3-17. (Use Table 3-9.)

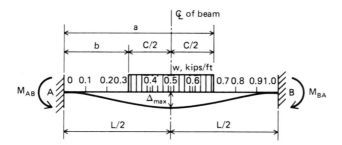

Fig. 3-17.

Solution

Use the same method as in Example 2-3. Subtract the partial, uniformly distributed load b from the partial, uniformly distributed load a to obtain the deflection of the partial, uniformly distributed load c.

When $c = 0.2L$:

$$\Delta_{0.4} = \frac{wL^4}{EI}(0.001751 - 0.000845) = \frac{wL^4}{EI} 0.000906 \text{ in downward}$$

$$\Delta_{0.5} = \frac{wL^4}{EI}(0.001804 - 0.000800) = \frac{wL^4}{EI} 0.001004 \text{ in downward} \qquad (ans.)$$

When $c = 0.4L$:

$$\Delta_{0.4} = \frac{wL^4}{EI}(0.002087 - 0.000429) = \frac{wL^4}{EI} 0.001658 \text{ in downward}$$

$$\Delta_{0.5} = \frac{wL^4}{EI}(0.002210 - 0.000394) = \frac{wL^4}{EI} 0.001816 \text{ in downward} \quad (ans.)$$

When $c = 0.6L$:

$$\Delta_{0.4} = \frac{wL^4}{EI}(0.002295 - 0.000149) = \frac{wL^4}{EI} 0.002146 \text{ in downward}$$

$$\Delta_{0.5} = \frac{wL^4}{EI}(0.002471 - 0.000133) = \frac{wL^4}{EI} 0.002338 \text{ in downward} \quad (ans.)$$

Since the beam is symmetrical about the centerline and the load is at the center point, the maximum deflection will occur at the center of the beam.

Example 3-8

Find the position of the approximate maximum deflection of a beam, fixed at both ends, carrying a partial, uniformly distributed load, as shown in Fig. 3-18. Let $c = 0.2L$, at the right end of the beam. (Use Table 3-9.)

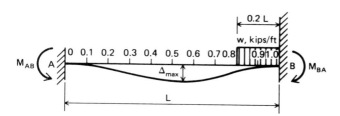

Fig. 3-18.

Solution

Try points 0.5 and 0.6 to find the approximate maximum deflection

$$\Delta_{0.5} = \frac{wL^4}{EI}(0.002604 - 0.002471) = \frac{wL^4}{EI} 0.000133 \text{ in downward}$$

$$\Delta_{0.6} = \frac{wL^4}{EI}(0.002400 - 0.002251) = \frac{wL^4}{EI} 0.000149 \text{ in downward}$$

In this case the largest deflection occurs at point 0.6. $\quad (ans.)$

3-5 COMBINING THE EQUATIONS OF DEFLECTION FOR LOADS AND END MOMENTS

Equations (2-3) and (2-18) give the deflections of a beam due to concentrated loads and uniformly distributed loads, respectively. Combining these equations with the deflection of the end moments, Eq. (3-12), we obtain

$$\Delta_x = \frac{L^3}{EI}\left[PC' + WC'' - \left(\frac{M_{AB}}{L}C_{AB} + \frac{M_{BA}}{L}C_{BA}\right)\right] \qquad (3\text{-}39)$$

where P = concentrated load
W = total, uniformly distributed load
C' = coefficient of deflection of a simple-support beam carrying a concentrated load (Table 2-1)
C'' = coefficient of deflection of a simple-support beam carrying a uniformly distributed load (Table 2-3)
C_{AB} = coefficient of deflection of a beam due to the end moment M_{AB} (Table 3-2)
C_{BA} = coefficient of deflection of a beam due to the end moment M_{BA} (Table 3-2)

This combined equation can be applied to a single-span beam with fixed-end moments or to a continuous beam with support moments.

TABLE 3-1 FORMULAS FOR THE FIXED-END MOMENTS

Type of loading	Beam fixed at both ends M_{AB}	Beam fixed at both ends M_{BA}	Right end hinged M_{AB}	Left end hinged M_{BA}
Concentrated load P at distance a from A, b from B	$\dfrac{Pab^2}{L^2}$	$\dfrac{Pa^2b}{L^2}$	$\dfrac{Pab}{2L^2}(L+b)$	$\dfrac{Pab}{2L^2}(L+a)$
Concentrated load P at midspan	$\dfrac{PL}{8}$	$\dfrac{PL}{8}$	$\dfrac{3PL}{16}$	$\dfrac{3PL}{16}$
Uniform load w over full span	$\dfrac{wL^2}{12}$	$\dfrac{wL^2}{12}$	$\dfrac{wL^2}{8}$	$\dfrac{wL^2}{8}$
Uniform load w over half span (from A)	$\dfrac{11}{192}wL^2$	$\dfrac{5}{192}wL^2$	$\dfrac{9}{128}wL^2$	$\dfrac{7}{128}wL^2$

TABLE 3-1 CONTINUED

Type of loading	Beam fixed at both ends		Right end hinged	Left end hinged
	M_{AB}	M_{BA}	M_{AB}	M_{BA}
(load near both ends, width mL each, intensity w)	$\dfrac{wL^2}{6} m^2(3 - 2m)$	$\dfrac{wL^2}{6} m^2(3 - 2m)$	$\dfrac{wL^2}{4} m^2(3 - 2m)$	$\dfrac{wL^2}{4} m^2(3 - 2m)$
(load at left end, width mL, intensity w)	$\dfrac{wL^2}{12} m^2(6 - 8m + 3m^2)$	$\dfrac{wL^2}{12} m^3(4 - 3m)$	$\dfrac{wL^2}{8} m^2(4 - 4m + m^2)$	$\dfrac{wL^2}{8} m^2(2 - m^2)$
(triangular load, max w at B)	$\dfrac{wL^2}{20}$	$\dfrac{wL^2}{30}$	$\dfrac{wL^2}{15}$	$\dfrac{7wL^2}{120}$
(triangular load, max w at midspan)	$\dfrac{23wL^2}{960}$	$\dfrac{7wL^2}{960}$	$\dfrac{53wL^2}{1,920}$	$\dfrac{37wL^2}{1,920}$

Beam				
(L/2, w, L/2 triangular load, fixed at B, pinned at A)	$\dfrac{wL^2}{30}$	$\dfrac{3wL^2}{160}$	$\dfrac{41wL^2}{960}$	$\dfrac{17wL^2}{480}$
(mL triangular load, fixed at B)	$\dfrac{wL^2}{60} m^2(10 - 10m + 3m^2)$	$\dfrac{wL^2}{60} m^3(5 - 3m)$	$\dfrac{wL^2}{120} m^2(20 - 15m + 3m^2)$	$\dfrac{wL^2}{120} m^2(10 - 3m^2)$
(mL triangular load reversed, fixed at B)	$\dfrac{wL^2}{30} m^2(10 - 15m + 6m^2)$	$\dfrac{wL^2}{20} m^3(5 - 4m)$	$\dfrac{wL^2}{120} (40 - 45m + 12m^2)$	

Note: M = fixed-end moment, kip · ft; P = concentrated load, kips; w = uniformly distributed load, kips/ft.

TABLE 3-2 COEFFICIENTS OF DEFLECTION OF A BEAM DUE TO END MOMENTS

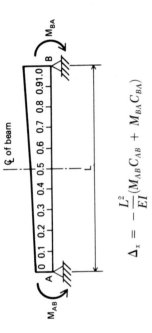

$$\Delta_x = -\frac{L^2}{EI}(M_{AB}C_{AB} + M_{BA}C_{BA})$$

	Coefficient C								
	$x/L = 0.1$	$x/L = 0.2$	$x/L = 0.3$	$x/L = 0.4$	$x/L = 0.5$	$x/L = 0.6$	$x/L = 0.7$	$x/L = 0.8$	$x/L = 0.9$
C_{AB}	0.02850	0.04800	0.05950	0.06400	0.06250	0.05600	0.04550	0.03200	0.01650
C_{BA}	0.01650	0.03200	0.04550	0.05600	0.06250	0.06400	0.05950	0.04800	0.02850
$C_{AB} + C_{BA}$*	0.04500	0.08000	0.10500	0.12000	0.12500	0.12000	0.10500	0.08000	0.04500

*When $M_{AB} = M_{BA}$.

TABLE 3-3 COEFFICIENTS OF DEFLECTION OF A BEAM, FIXED AT ONE END AND SUPPORTED AT THE OTHER, CARRYING A CONCENTRATED LOAD

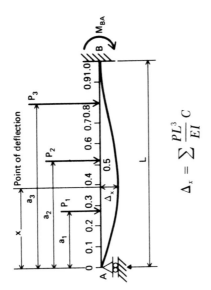

$$\Delta_x = \sum \frac{PL^3}{EI} C$$

Position of concentrated load, a/L	Coefficient C								
	$x/L = 0.1$	$x/L = 0.2$	$x/L = 0.3$	$x/L = 0.4$	$x/L = 0.5$	$x/L = 0.6$	$x/L = 0.7$	$x/L = 0.8$	$x/L = 0.9$
0.025	0.000504	0.000798	0.000917	0.000899	0.000780	0.000599	0.000393	0.000200	0.000056
0.050	0.000995	0.001585	0.001826	0.001791	0.001556	0.001196	0.000785	0.000399	0.000112
0.075	0.001459	0.002350	0.002717	0.002670	0.002322	0.001785	0.001173	0.000596	0.000168
0.100	0.001883	0.003083	0.003581	0.003528	0.003073	0.002365	0.001555	0.000791	0.000223
0.125	0.002257	0.003771	0.004410	0.004359	0.003805	0.002932	0.001929	0.000982	0.000277
0.150	0.002580	0.004404	0.005196	0.005157	0.004512	0.003483	0.002294	0.001168	0.000329
0.175	0.002854	0.004971	0.005928	0.005914	0.005190	0.004014	0.002648	0.001350	0.000381
0.200	0.003083	0.005461	0.006599	0.006624	0.005833	0.004523	0.002988	0.001525	0.000431
0.225	0.003267	0.005866	0.007199	0.007280	0.006438	0.005005	0.003313	0.001694	0.000479
0.250	0.003410	0.006187	0.007720	0.007875	0.006999	0.005458	0.003621	0.001854	0.000525
0.275	0.003514	0.006430	0.008153	0.008403	0.007511	0.005879	0.003910	0.002006	0.000568
0.300	0.003581	0.006599	0.008489	0.008856	0.007969	0.006264	0.004178	0.002148	0.000610

0.325	0.003614	0.006698	0.008722	0.009228	0.008368	0.006610	0.004424	0.002280	0.000648
0.350	0.003614	0.006732	0.008857	0.009513	0.008704	0.006914	0.004644	0.002400	0.000684
0.375	0.003585	0.006706	0.008899	0.009703	0.008972	0.007172	0.004838	0.002508	0.000716
0.400	0.003528	0.006624	0.008856	0.009792	0.009167	0.007381	0.005004	0.002603	0.000745
0.425	0.003446	0.006491	0.008735	0.009775	0.009283	0.007539	0.005139	0.002684	0.000771
0.450	0.003341	0.006312	0.008542	0.009660	0.009316	0.007641	0.005242	0.002749	0.000792
0.475	0.003216	0.006091	0.008284	0.009454	0.009262	0.007685	0.005311	0.002800	0.000810
0.500	0.003073	0.005833	0.007969	0.009167	0.009115	0.007667	0.005344	0.002833	0.000823
0.525	0.002914	0.005543	0.007602	0.008807	0.008872	0.007584	0.005339	0.002849	0.000832
0.550	0.002741	0.005224	0.007191	0.008384	0.008543	0.007432	0.005293	0.002847	0.000835
0.575	0.002558	0.004883	0.006743	0.007905	0.008138	0.007210	0.005207	0.002826	0.000834
0.600	0.002365	0.004523	0.006264	0.007381	0.007667	0.006912	0.005076	0.002784	0.000828
0.625	0.002167	0.004148	0.005761	0.006820	0.007141	0.006539	0.004900	0.002721	0.000816
0.650	0.001964	0.003765	0.005241	0.006231	0.006572	0.006101	0.004676	0.002637	0.000799
0.675	0.001759	0.003376	0.004712	0.005623	0.005969	0.005609	0.004403	0.002530	0.000776
0.700	0.001555	0.002988	0.004178	0.005004	0.005344	0.005076	0.004079	0.002399	0.000746
0.725	0.001354	0.002604	0.003648	0.004384	0.004707	0.004515	0.003705	0.002243	0.000710
0.750	0.001158	0.002229	0.003129	0.003771	0.004069	0.003938	0.003290	0.002062	0.000668
0.775	0.000969	0.001868	0.002626	0.003174	0.003441	0.003356	0.002851	0.001855	0.000619
0.800	0.000791	0.001525	0.002148	0.002603	0.002833	0.002784	0.002399	0.001621	0.000563
0.825	0.000624	0.001206	0.001700	0.002065	0.002257	0.002233	0.001949	0.001362	0.000499
0.850	0.000473	0.000913	0.001290	0.001570	0.001723	0.001714	0.001514	0.001089	0.000428
0.875	0.000338	0.000654	0.000924	0.001128	0.001241	0.001242	0.001109	0.000818	0.000350
0.900	0.000223	0.000431	0.000610	0.000745	0.000823	0.000828	0.000746	0.000563	0.000263
0.925	0.000129	0.000249	0.000353	0.000433	0.000479	0.000484	0.000440	0.000339	0.000171
0.950	0.000059	0.000114	0.000162	0.000198	0.000220	0.000224	0.000205	0.000160	0.000086
0.975	0.000015	0.000029	0.000042	0.000051	0.000057	0.000058	0.000053	0.000043	0.000024
1.000	0.000000	0.000000	0.000000	0.000000	0.000000	0.000000	0.000000	0.000000	0.000000

Note: The boxed areas contain the largest coefficient of deflection for each load position.

TABLE 3-4 MAXIMUM COEFFICIENTS OF DEFLECTION OF A BEAM, FIXED AT ONE END AND SUPPORTED AT THE OTHER, CARRYING A CONCENTRATED LOAD

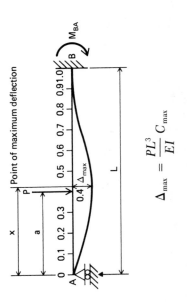

$$\Delta_{max} = \frac{PL^3}{EI} C_{max}$$

Position of concentrated load, a/L	Maximum coefficient, C_{max}	Position of max deflection, x/L	Position of concentrated load, a/L	Maximum coefficient, C_{max}	Position of max deflection, x/L
0.025	0.000925	0.334	0.525	0.009002	0.456
0.050	0.001841	0.334	0.550	0.008621	0.464
0.075	0.002741	0.336	0.575	0.008180	0.473
0.100	0.003618	0.338	0.600	0.007686	0.480
0.125	0.004462	0.340	0.625	0.007148	0.488
0.150	0.005268	0.343	0.650	0.006573	0.495
0.175	0.006026	0.347	0.675	0.005969	0.502
0.200	0.006732	0.351	0.700	0.005346	0.509
0.225	0.007378	0.356	0.725	0.004713	0.516
0.250	0.007958	0.362	0.750	0.004080	0.522
0.275	0.008466	0.368	0.775	0.003456	0.528
0.300	0.008899	0.375	0.800	0.002851	0.535
0.325	0.009251	0.382	0.825	0.002276	0.540
0.350	0.009521	0.390	0.850	0.001741	0.546
0.375	0.009703	0.399	0.875	0.001257	0.552
0.400	0.009798	0.408	0.900	0.000836	0.557
0.425	0.009804	0.419	0.925	0.000488	0.562
0.450	0.009723	0.429	0.950	0.000225	0.567
0.475	0.009559	0.438	0.975	0.000058	0.572
0.500	0.009317	0.447	1.000	0.000000	0.000

TABLE 3-5 COEFFICIENTS OF DEFLECTION OF A BEAM, FIXED AT ONE END AND SUPPORTED AT THE OTHER, CARRYING A UNIFORMLY DISTRIBUTED LOAD EXTENDING FROM ONE END

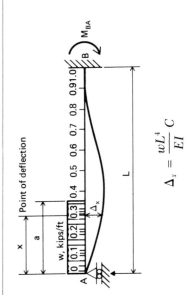

$$\Delta_x = \frac{wL^4}{EI} C$$

Length of uniform load extending from left support, a/L	$x/L = 0.1$	$x/L = 0.2$	$x/L = 0.3$	$x/L = 0.4$	Coefficient C $x/L = 0.5$	$x/L = 0.6$	$x/L = 0.7$	$x/L = 0.8$	$x/L = 0.9$
0.025	0.000006	0.000010	0.000011	0.000011	0.000010	0.000007	0.000005	0.000002	0.000001
0.050	0.000025	0.000040	0.000046	0.000045	0.000039	0.000030	0.000020	0.000010	0.000003
0.075	0.000056	0.000089	0.000103	0.000101	0.000087	0.000067	0.000044	0.000022	0.000006
0.100	0.000098	0.000157	0.000181	0.000178	0.000155	0.000119	0.000078	0.000040	0.000011
0.125	0.000150	0.000243	0.000281	0.000277	0.000241	0.000185	0.000122	0.000062	0.000017
0.150	0.000210	0.000345	0.000402	0.000396	0.000345	0.000266	0.000175	0.000089	0.000025
0.175	0.000278	0.000462	0.000541	0.000534	0.000466	0.000359	0.000236	0.000120	0.000034
0.200	0.000352	0.000593	0.000697	0.000691	0.000604	0.000466	0.000307	0.000156	0.000044

0.225	0.000432	0.000735	0.000870	0.000865	0.000758	0.000585	0.000386	0.000197	0.000055
0.250	0.000515	0.000886	0.001057	0.001055	0.000928	0.000716	0.000472	0.000241	0.000068
0.275	0.000602	0.001044	0.001255	0.001258	0.001107	0.000858	0.000567	0.000289	0.000082
0.300	0.000691	0.001207	0.001464	0.001474	0.001301	0.001010	0.000668	0.000341	0.000096
0.325	0.000781	0.001373	0.001679	0.001700	0.001505	0.001171	0.000775	0.000396	0.000112
0.350	0.000871	0.001541	0.001899	0.001935	0.001719	0.001340	0.000889	0.000455	0.000129
0.375	0.000961	0.001709	0.002121	0.002175	0.001940	0.001516	0.001007	0.000516	0.000146
0.400	0.001050	0.001876	0.002343	0.002419	0.002167	0.001698	0.001130	0.000580	0.000165
0.425	0.001138	0.002040	0.002563	0.002664	0.002397	0.001885	0.001257	0.000646	0.000183
0.450	0.001222	0.002200	0.002779	0.002907	0.002630	0.002075	0.001387	0.000714	0.000203
0.475	0.001304	0.002355	0.002990	0.003146	0.002863	0.002266	0.001519	0.000784	0.000223
0.500	0.001383	0.002504	0.003193	0.003379	0.003092	0.002458	0.001652	0.000854	0.000243
0.525	0.001458	0.002646	0.003388	0.003604	0.003317	0.002649	0.001786	0.000925	0.000264
0.550	0.001529	0.002781	0.003573	0.003819	0.003535	0.002837	0.001919	0.000996	0.000285
0.575	0.001595	0.002907	0.003747	0.004023	0.003744	0.003020	0.002050	0.001067	0.000306
0.600	0.001656	0.003025	0.003910	0.004214	0.003942	0.003197	0.002179	0.001138	0.000327
0.625	0.001713	0.003133	0.004060	0.004391	0.004127	0.003365	0.002304	0.001206	0.000347
0.650	0.001765	0.003232	0.004198	0.004555	0.004298	0.003523	0.002423	0.001273	0.000367
0.675	0.001811	0.003322	0.004322	0.004703	0.004455	0.003670	0.002537	0.001338	0.000387
0.700	0.001853	0.003401	0.004433	0.004836	0.004597	0.003803	0.002643	0.001400	0.000406
0.725	0.001889	0.003471	0.004531	0.004953	0.004722	0.003923	0.002741	0.001458	0.000424
0.750	0.001920	0.003532	0.004616	0.005055	0.004832	0.004029	0.002828	0.001512	0.000442
0.775	0.001947	0.003583	0.004687	0.005142	0.004926	0.004120	0.002905	0.001561	0.000458
0.800	0.001969	0.003625	0.004747	0.005214	0.005004	0.004197	0.002971	0.001604	0.000473
0.825	0.001987	0.003659	0.004795	0.005272	0.005068	0.004259	0.003025	0.001642	0.000486
0.850	0.002000	0.003686	0.004832	0.005317	0.005117	0.004309	0.003068	0.001672	0.000497
0.875	0.002010	0.003705	0.004860	0.005351	0.005154	0.004346	0.003101	0.001696	0.000507
0.900	0.002017	0.003719	0.004879	0.005374	0.005180	0.004371	0.003124	0.001713	0.000515
0.925	0.002022	0.003727	0.004891	0.005389	0.005196	0.004388	0.003139	0.001724	0.000520
0.950	0.002024	0.003731	0.004897	0.005397	0.005205	0.004396	0.003147	0.001731	0.000523
0.975	0.002025	0.003733	0.004900	0.005400	0.005208	0.004400	0.003150	0.001733	0.000525
1.000	0.002025	0.003733	0.004900	0.005400	0.005208	0.004400	0.003150	0.001733	0.000525

Note: The boxed areas contain the largest coefficient of deflection for each load position.

TABLE 3-6 MAXIMUM COEFFICIENTS OF DEFLECTION OF A BEAM, FIXED AT ONE END AND SUPPORTED AT THE OTHER, CARRYING A UNIFORMLY DISTRIBUTED LOAD EXTENDING FROM ONE END

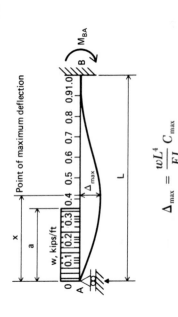

$$\Delta_{max} = \frac{wL^4}{EI} C_{max}$$

Length of uniform load extending from left support, a/L	Maximum coefficient, C_{max}	Position of max deflection, x/L	Length of uniform load extending from left support, a/L	Maximum coefficient, C_{max}	Position of max deflection, x/L
0.025	0.000012	0.333	0.525	0.003606	0.391
0.050	0.000046	0.334	0.550	0.003820	0.395
0.075	0.000104	0.335	0.575	0.004023	0.398
0.100	0.000183	0.336	0.600	0.004214	0.401
0.125	0.000284	0.337	0.625	0.004392	0.404
0.150	0.000406	0.338	0.650	0.004556	0.407
0.175	0.000547	0.340	0.675	0.004706	0.409
0.200	0.000707	0.342	0.700	0.004840	0.412
0.225	0.000883	0.345	0.725	0.004959	0.414
0.250	0.001075	0.347	0.750	0.005063	0.415
0.275	0.001280	0.350	0.775	0.005151	0.417
0.300	0.001496	0.354	0.800	0.005225	0.418
0.325	0.001722	0.357	0.825	0.005285	0.419
0.350	0.001956	0.361	0.850	0.005331	0.420
0.375	0.002194	0.365	0.875	0.005366	0.421
0.400	0.002435	0.370	0.900	0.005390	0.421
0.425	0.002677	0.374	0.925	0.005405	0.421
0.450	0.002917	0.379	0.950	0.005413	0.421
0.475	0.003153	0.383	0.975	0.005415	0.421
0.500	0.003383	0.387	1.000	0.005416	0.422

TABLE 3-7 COEFFICIENTS OF DEFLECTION OF A BEAM, FIXED AT BOTH ENDS, CARRYING A CONCENTRATED LOAD

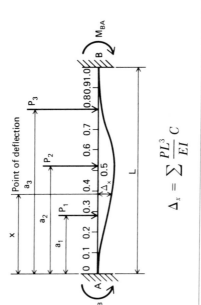

$$\Delta_x = \sum \frac{PL^3}{EI} C$$

Position of concentrated load, a/L	Coefficient C								
	$x/L = 0.1$	$x/L = 0.2$	$x/L = 0.3$	$x/L = 0.4$	$x/L = 0.5$	$x/L = 0.6$	$x/L = 0.7$	$x/L = 0.8$	$x/L = 0.9$
0.025	0.000023	0.000038	0.000044	0.000043	0.000038	0.000029	0.000019	0.000010	0.000003
0.050	0.000081	0.000141	0.000167	0.000167	0.000146	0.000113	0.000074	0.000038	0.000011
0.075	0.000159	0.000297	0.000358	0.000359	0.000316	0.000245	0.000162	0.000083	0.000023
0.100	0.000243	0.000491	0.000604	0.000612	0.000542	0.000421	0.000279	0.000143	0.000040
0.125	0.000319	0.000708	0.000893	0.000914	0.000814	0.000635	0.000422	0.000216	0.000061
0.150	0.000385	0.000936	0.001213	0.001256	0.001125	0.000882	0.000587	0.000301	0.000086
0.175	0.000442	0.001160	0.001551	0.001626	0.001467	0.001156	0.000772	0.000397	0.000113
0.200	0.000491	0.001365	0.001895	0.002016	0.001833	0.001451	0.000972	0.000501	0.000143
0.225	0.000531	0.001542	0.002233	0.002415	0.002215	0.001762	0.001185	0.000613	0.000175
0.250	0.000563	0.001688	0.002552	0.002813	0.002604	0.002083	0.001406	0.000729	0.000208
0.275	0.000587	0.001805	0.002841	0.003199	0.002993	0.002410	0.001634	0.000850	0.000243
0.300	0.000604	0.001895	0.003087	0.003564	0.003375	0.002736	0.001863	0.000972	0.000279

0.325	0.000615	0.001959	0.003281	0.003898	0.003741	0.003056	0.002091	0.001095	0.000315
0.350	0.000620	0.002000	0.003422	0.004189	0.004083	0.003365	0.002315	0.001217	0.000351
0.375	0.000618	0.002018	0.003516	0.004430	0.004395	0.003656	0.002531	0.001336	0.000387
0.400	0.000612	0.002016	0.003564	0.004608	0.004667	0.003925	0.002736	0.001451	0.000421
0.425	0.000601	0.001995	0.003571	0.004717	0.004892	0.004166	0.002926	0.001559	0.000455
0.450	0.000585	0.001956	0.003539	0.004759	0.005062	0.004374	0.003098	0.001660	0.000486
0.475	0.000565	0.001902	0.003473	0.004741	0.005171	0.004543	0.003249	0.001752	0.000515
0.500	0.000542	0.001833	0.003375	0.004667	0.005208	0.004667	0.003375	0.001833	0.000542
0.525	0.000515	0.001752	0.003249	0.004543	0.005171	0.004741	0.003473	0.001902	0.000565
0.550	0.000486	0.001660	0.003098	0.004374	0.005062	0.004759	0.003539	0.001956	0.000585
0.575	0.000455	0.001559	0.002926	0.004166	0.004892	0.004717	0.003571	0.001995	0.000601
0.600	0.000421	0.001451	0.002736	0.003925	0.004667	0.004608	0.003564	0.002016	0.000612
0.625	0.000387	0.001336	0.002531	0.003656	0.004395	0.004430	0.003516	0.002018	0.000618
0.650	0.000351	0.001217	0.002315	0.003365	0.004083	0.004189	0.003422	0.002000	0.000620
0.675	0.000315	0.001095	0.002091	0.003056	0.003741	0.003898	0.003281	0.001959	0.000615
0.700	0.000279	0.000972	0.001863	0.002736	0.003375	0.003564	0.003087	0.001895	0.000604
0.725	0.000243	0.000850	0.001634	0.002410	0.002993	0.003199	0.002841	0.001805	0.000587
0.750	0.000208	0.000729	0.001406	0.002083	0.002604	0.002813	0.002552	0.001688	0.000563
0.775	0.000175	0.000613	0.001185	0.001762	0.002215	0.002415	0.002233	0.001542	0.000531
0.800	0.000143	0.000501	0.000972	0.001451	0.001833	0.002016	0.001895	0.001365	0.000491
0.825	0.000113	0.000397	0.000772	0.001156	0.001467	0.001626	0.001551	0.001160	0.000442
0.850	0.000086	0.000301	0.000587	0.000882	0.001125	0.001256	0.001213	0.000936	0.000385
0.875	0.000061	0.000216	0.000422	0.000635	0.000814	0.000914	0.000893	0.000708	0.000319
0.900	0.000040	0.000143	0.000279	0.000421	0.000542	0.000612	0.000604	0.000491	0.000243
0.925	0.000023	0.000083	0.000162	0.000245	0.000316	0.000359	0.000358	0.000297	0.000159
0.950	0.000011	0.000038	0.000074	0.000113	0.000146	0.000167	0.000167	0.000141	0.000081
0.975	0.000003	0.000010	0.000019	0.000029	0.000038	0.000043	0.000044	0.000038	0.000023
1.000	0.000000	0.000000	0.000000	0.000000	0.000000	0.000000	0.000000	0.000000	0.000000

Note: The boxed areas contain the largest coefficient of deflection for each load position.

TABLE 3-8 MAXIMUM COEFFICIENTS OF DEFLECTION OF A BEAM, FIXED AT BOTH ENDS, CARRYING A CONCENTRATED LOAD

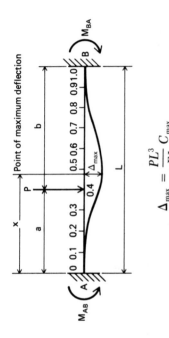

$$\Delta_{max} = \frac{PL^3}{EI} C_{max}$$

74

Position of concentrated load, a/L	Maximum coefficient, C_{max}	Position of max deflection, x/L	Position of concentrated load, a/L	Maximum coefficient, C_{max}	Position of max deflection, x/L
0.025	0.000044	0.339	0.525	0.005179	0.512
0.050	0.000170	0.345	0.550	0.005093	0.524
0.075	0.000365	0.351	0.575	0.004952	0.535
0.100	0.000620	0.357	0.600	0.004760	0.545
0.125	0.000923	0.364	0.625	0.004521	0.556
0.150	0.001264	0.370	0.650	0.004240	0.565
0.175	0.001633	0.377	0.675	0.003922	0.574
0.200	0.002020	0.385	0.700	0.003573	0.583
0.225	0.002416	0.392	0.725	0.003201	0.592
0.250	0.002813	0.400	0.750	0.002813	0.600
0.275	0.003201	0.408	0.775	0.002416	0.608
0.300	0.003573	0.417	0.800	0.002020	0.615
0.325	0.003922	0.426	0.825	0.001633	0.623
0.350	0.004240	0.435	0.850	0.001264	0.630
0.375	0.004521	0.444	0.875	0.000923	0.636
0.400	0.004760	0.455	0.900	0.000620	0.643
0.425	0.004952	0.465	0.925	0.000365	0.649
0.450	0.005093	0.476	0.950	0.000170	0.655
0.475	0.005179	0.488	0.975	0.000044	0.661
0.500	0.005208	0.500	1.000	0.000000	0.000

TABLE 3-9 COEFFICIENTS OF DEFLECTION OF A BEAM, FIXED AT BOTH ENDS, CARRYING A UNIFORMLY DISTRIBUTED LOAD EXTENDING FROM ONE END

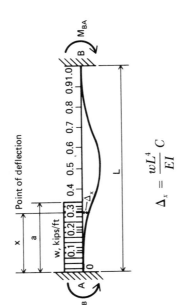

$$\Delta_x = \frac{wL^4}{EI} C$$

Length of uniform load extending from left support, a/L	Coefficient C								
	$x/L = 0.1$	$x/L = 0.2$	$x/L = 0.3$	$x/L = 0.4$	$x/L = 0.5$	$x/L = 0.6$	$x/L = 0.7$	$x/L = 0.8$	$x/L = 0.9$
0.025	0.000000	0.000000	0.000000	0.000000	0.000000	0.000000	0.000000	0.000000	0.000000
0.050	0.000001	0.000002	0.000003	0.000003	0.000002	0.000002	0.000001	0.000001	0.000000
0.075	0.000004	0.000008	0.000009	0.000009	0.000008	0.000006	0.000004	0.000002	0.000001
0.100	0.000009	0.000018	0.000021	0.000021	0.000019	0.000015	0.000010	0.000005	0.000001
0.125	0.000016	0.000033	0.000040	0.000040	0.000036	0.000028	0.000018	0.000009	0.000003
0.150	0.000025	0.000053	0.000066	0.000067	0.000060	0.000047	0.000031	0.000016	0.000004
0.175	0.000036	0.000079	0.000101	0.000103	0.000092	0.000072	0.000048	0.000025	0.000007
0.200	0.000047	0.000111	0.000144	0.000149	0.000133	0.000105	0.000070	0.000036	0.000010
0.225	0.000060	0.000147	0.000195	0.000204	0.000184	0.000145	0.000097	0.000050	0.000014
0.250	0.000074	0.000188	0.000255	0.000270	0.000244	0.000193	0.000129	0.000066	0.000019
0.275	0.000088	0.000231	0.000323	0.000345	0.000314	0.000249	0.000167	0.000086	0.000025
0.300	0.000103	0.000278	0.000397	0.000429	0.000394	0.000313	0.000211	0.000109	0.000031

0.325	0.000118	0.000326	0.000477	0.000523	0.000483	0.000386	0.000260	0.000135	0.000038
0.350	0.000134	0.000376	0.000560	0.000624	0.000581	0.000466	0.000315	0.000164	0.000047
0.375	0.000149	0.000426	0.000647	0.000732	0.000687	0.000554	0.000376	0.000196	0.000056
0.400	0.000165	0.000476	0.000736	0.000845	0.000800	0.000649	0.000442	0.000230	0.000066
0.425	0.000180	0.000526	0.000825	0.000962	0.000920	0.000750	0.000512	0.000268	0.000077
0.450	0.000195	0.000576	0.000914	0.001080	0.001044	0.000857	0.000588	0.000308	0.000089
0.475	0.000209	0.000624	0.001002	0.001199	0.001172	0.000968	0.000667	0.000351	0.000101
0.500	0.000223	0.000671	0.001088	0.001317	0.001302	0.001083	0.000750	0.000396	0.000115
0.525	0.000236	0.000716	0.001170	0.001432	0.001432	0.001201	0.000836	0.000443	0.000128
0.550	0.000249	0.000758	0.001250	0.001543	0.001560	0.001320	0.000923	0.000491	0.000143
0.575	0.000260	0.000799	0.001325	0.001650	0.001685	0.001438	0.001012	0.000540	0.000158
0.600	0.000271	0.000836	0.001396	0.001751	0.001804	0.001555	0.001102	0.000590	0.000173
0.625	0.000281	0.000871	0.001462	0.001846	0.001918	0.001668	0.001190	0.000641	0.000188
0.650	0.000291	0.000903	0.001522	0.001934	0.002024	0.001776	0.001277	0.000691	0.000204
0.675	0.000299	0.000932	0.001577	0.002014	0.002121	0.001877	0.001361	0.000741	0.000219
0.700	0.000306	0.000958	0.001627	0.002087	0.002210	0.001971	0.001441	0.000789	0.000234
0.725	0.000313	0.000981	0.001671	0.002151	0.002290	0.002055	0.001515	0.000835	0.000249
0.750	0.000319	0.001000	0.001709	0.002207	0.002360	0.002130	0.001582	0.000879	0.000264
0.775	0.000323	0.001017	0.001741	0.002255	0.002420	0.002196	0.001642	0.000919	0.000277
0.800	0.000327	0.001031	0.001768	0.002295	0.002471	0.002251	0.001694	0.000956	0.000290
0.825	0.000331	0.001042	0.001790	0.002328	0.002512	0.002297	0.001737	0.000987	0.000302
0.850	0.000333	0.001051	0.001807	0.002353	0.002544	0.002333	0.001771	0.001014	0.000312
0.875	0.000335	0.001057	0.001819	0.002372	0.002569	0.002360	0.001798	0.001034	0.000321
0.900	0.000336	0.001062	0.001828	0.002385	0.002585	0.002379	0.001816	0.001049	0.000328
0.925	0.000337	0.001065	0.001833	0.002394	0.002596	0.002391	0.001828	0.001059	0.000333
0.950	0.000337	0.001066	0.001836	0.002398	0.002602	0.002397	0.001835	0.001064	0.000336
0.975	0.000337	0.001067	0.001837	0.002400	0.002604	0.002400	0.001837	0.001066	0.000337
1.000	0.000338	0.001067	0.001838	0.002400	0.002604	0.002400	0.001838	0.001067	0.000338

Note: The boxed areas contain the largest coefficient of deflection for each load position.

TABLE 3-10 MAXIMUM COEFFICIENTS OF DEFLECTION OF A BEAM, FIXED AT BOTH ENDS, CARRYING A UNIFORMLY DISTRIBUTED LOAD EXTENDING FROM ONE END

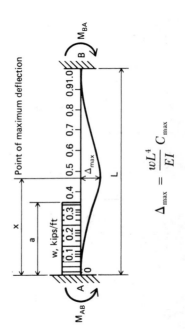

$$\Delta_{max} = \frac{wL^4}{EI} C_{max}$$

Length of uniform load extending from left support, a/L	Maximum coefficient, C_{max}	Position of max deflection, x/L	Length of uniform load extending from left support, a/L	Maximum coefficient, C_{max}	Position of max deflection, x/L
0.025	0.000001	0.337	0.525	0.001465	0.450
0.050	0.000003	0.342	0.550	0.001588	0.456
0.075	0.000010	0.346	0.575	0.001707	0.461
0.100	0.000022	0.351	0.600	0.001822	0.466
0.125	0.000041	0.356	0.625	0.001931	0.471
0.150	0.000068	0.360	0.650	0.002034	0.476
0.175	0.000104	0.365	0.675	0.002129	0.480
0.200	0.000150	0.370	0.700	0.002216	0.483
0.225	0.000205	0.376	0.725	0.002293	0.487
0.250	0.000270	0.381	0.750	0.002362	0.490
0.275	0.000345	0.386	0.775	0.002422	0.492
0.300	0.000430	0.392	0.800	0.002472	0.494
0.325	0.000523	0.398	0.825	0.002512	0.496
0.350	0.000624	0.404	0.850	0.002545	0.497
0.375	0.000732	0.410	0.875	0.002569	0.498
0.400	0.000847	0.417	0.900	0.002585	0.499
0.425	0.000966	0.423	0.925	0.002596	0.500
0.450	0.001089	0.430	0.950	0.002602	0.500
0.475	0.001214	0.437	0.975	0.002604	0.500
0.500	0.001340	0.443	1.000	0.002604	0.500

4
Deflection of Continuous Beams

SUPPORT MOMENTS

In a continuous beam, we have positive moments between the supports and negative moments at the supports. The negative moments at the supports we call the *support moments*. These counteract the positive moments under loads occurring between the supports. Therefore, by using a continuous beam in the design, the amount of deflection at the center of the beam can be reduced.

The procedure for determining the deflection of each span of a continuous beam is similar to that for determining the deflection of a single-span beam restrained at one or both ends. The only difference is that in the continuous beam the support moments take the place of the fixed-end moments. In calculating the deflection of each span of a continuous beam, the support moments must be found before the deflection of the span can be calculated.

The continuous beam is usually a statically indeterminate structure. The higher the degree of indetermination, the more time-consuming the solution of the problem. Several methods generally employed to find the support moments for a continuous beam are introduced here.

THREE-MOMENT EQUATION METHOD

The three-moment equation expresses the relationship between the bending moments at three successive supports of a continuous beam that is subjected to certain applied loading on the various spans, with or without unequal settlements of supports. This relationship can be derived on the basis of the continuity of the elastic curve over the middle support; that is, the slope of the tangent at the right end of the left span must be equal to the slope of the tangent at the left end of the right span. The equation is given as follows. It can be used to solve statically indeterminate beams to any degree.

$$\frac{M_A L_1}{I_1} + 2M_B\left(\frac{L_1}{I_1} + \frac{L_2}{I_2}\right) + \frac{M_C L_2}{I_2} = -\frac{w_1 L_1^3}{4I_1} - \frac{w_2 L_2^3}{4I_2}$$

$$-\frac{1}{I_1}\sum P_1 L_1^2(k_1 - k_1^3) - \frac{1}{I_2}\sum P_2 L_2^2(k_2 - k_2^3) + 6E\left(\frac{m}{L_1} + \frac{n}{L_2}\right) \quad (4\text{-}1)$$

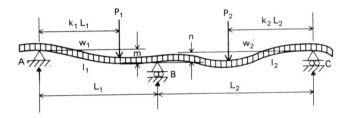

Fig. 4-1.

In Eq. (4-1) m is positive when the support B is depressed below A, and n is positive when the support B is depressed below C (see Fig. 4-1).

Example 4-1

Determine the Δ_D and Δ_E for a continuous beam, fixed at end A and supported at B and C (see Fig. 4-2). The quantity EI is constant. Use the three-moment equation method to determine the deflections.

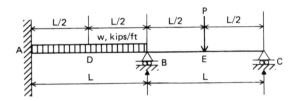

Fig. 4-2.

Solution

When the beam has a fixed end, an imaginary span of zero length is added to the fixed-end span. Thus,

for span AB $\quad 0 + 2M_A(L + 0) + M_B L = -\frac{1}{4}wL^3$

and for spans AB and BC

$$M_A L + 2M_B(L + L) = -\frac{1}{4}wL^3 - PL^2(0.5 - 0.5^3)$$

Solving these equations simultaneously, we get

$$M_A = -\frac{3wL^2}{28} + \frac{1.5PL}{28}$$

$$M_B = -\frac{wL^2}{28} - \frac{3PL}{28}$$

With these support moments determined, the deflection of the beam can be found.

The deflection at D due to a uniformly distributed load with $C'' = 0.013021$ (from Table 2-3) and $C_{AB} = C_{BA} = 0.0625$ (from Table 3-2) is, by Eq. (3-39),

$$\Delta_D = \frac{L^3}{EI}\left(0.013021wL - 0.0625\,\frac{4wL + 1.5P}{28}\right)$$

$$= \frac{L^3}{EI}(0.004093wL - 0.003348P) \quad \text{in downward} \quad (ans.)$$

The deflection at E due to a concentrated load with $C' = 0.020833$ (from Table 2-1) and $C_{BC} = 0.0625$ (from Table 3-2) is, by Eq. (3-39),

$$\Delta_E = \frac{L^3}{EI}\left(0.020833P - 0.625\,\frac{wL + 3P}{28}\right)$$

$$= \frac{L^3}{EI}(0.014137P - 0.002232wL) \quad \text{in downward} \quad (ans.)$$

Example 4-2

A two-span continuous beam is designed with two different cross sections in the spans as shown in Fig. 4-3. Determine Δ_D and Δ_E, using the three-moment equation method.

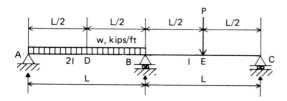

Fig. 4-3.

Solution

For spans AB and BC

$$0 + 2M_B\left(\frac{L}{2I} + \frac{L}{I}\right) + 0 = -\frac{1}{4}\frac{wL^3}{2I} - \frac{1}{I}PL^2(0.5 - 0.5^3)$$

Therefore

$$M_B = -\frac{wL^2}{24} - \frac{PL}{8}$$

The deflection at D due to a uniformly distributed load with $C'' = 0.013021$ (from Table 2-3) and $C_{BA} = 0.0625$ (from Table 3-2) is, by Eq. (3-39),

$$\Delta_D = \frac{L^3}{EI}\left(0.013021wL - 0.0625\,\frac{wL + 3P}{24}\right)$$

$$= \frac{L^3}{EI}(0.010417wL - 0.007813P) \qquad \text{in downward} \qquad (ans.)$$

The deflection at E due to a concentrated load with $C' = 0.020833$ (from Table 2-1) and $C_{BC} = 0.0625$ (from Table 3-2) is, by Eq. (3-39),

$$\Delta_E = \frac{L^3}{EI}\left(0.020833P - 0.0625\,\frac{wL + 3P}{24}\right)$$

$$= \frac{L^3}{EI}(0.013021P - 0.002604wL) \qquad \text{in downward} \qquad (ans.)$$

4-3 CONSISTENT DEFLECTION METHOD

The consistent deflection method is the most general method for solving statically indeterminate structurals, beams, frames, and trusses. The basic solution involves the fact that any number of indeterminate members can be considered as unknown forces. Each unknown force is called a *redundant*, and each redundant is considered as having a load acting on it.

Thus, the statically indeterminate structure can now be solved as a statically determinate structure. By relating the deflection Δ_x from the applied load (redundant removed) and the deflection δ_x from a unit load applied at the point of redundant reaction, we can find the redundant force R_x

$$R_x = \frac{\Delta_x}{\delta_x} \qquad (4\text{-}2)$$

This work can be reduced or simplified by the application of Maxwell's law of reciprocal deflections.[1]

$$\delta_{12} = \delta_{21} \qquad (4\text{-}3)$$

where δ_{12} = deflection of point 1 due to a unit load applied at point 2
δ_{21} = deflection of point 2 due to a unit load applied at point 1

Maxwell's law is perfectly general and is applicable to any indeterminate structure for finding the moments and shear at each support.

---
[1] C. H. Norris and J. B. Wilbur, *Elementary Structural Analysis*, 2d ed., McGraw-Hill, New York, 1960, p. 389.

Example 4-3

A continuous beam with two equal spans is loaded with two concentrated loads P at the center of each span, as is shown in Fig. 4-4. Determine the support moments at B and the deflection at D, using the consistent deflection method.

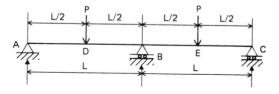

Fig. 4-4.

Solution

Take R_B as a redundant and use Table 2-1 to find Δ_B, the deflection at B due to two equal concentrated loads P (R_B removed), and δ_B, the deflection at B due to the redundant reaction R_B.

$$\Delta_B = \frac{P(2L)^3}{EI} 2 \times 0.014323 = \frac{PL^3}{EI} 0.229168$$

$$\delta_B = \frac{1(2L)^3}{EI} 0.020833 = \frac{L^3}{EI} 0.166664$$

By Eq. (4-2)
$$R_B = \frac{11P}{8}$$

$$R_A = R_C = \frac{5P}{16}$$

$$M_{BA} = -\frac{3PL}{16}$$

$$M_{BC} = +\frac{3PL}{16}$$
(ans.)

The deflection at D due to the concentrated load with $C' = 0.020833$ (from Table 2-1) and $C_{BA} = 0.0625$ (from Table 3-2) is, by Eq. (3-39),

$$\Delta_D = \frac{L^3}{EI}\left(0.020833P - 0.0625 \times \frac{3}{16}P\right)$$

$$= \frac{PL^3}{EI} 0.009114 \quad \text{in downward} \quad \text{(ans.)}$$

Example 4-4

A continuous beam with two equal spans is loaded with a uniformly distributed load over the whole of spans 1 and 2 as shown in Fig. 4-5. Determine the support moments at B and the deflection at D, using the consistent deflection method.

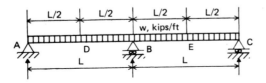

Fig. 4-5.

Solution

Take R_B as a redundant and use Table 2-3 to find Δ_B, the deflection at B due to the uniformly distributed load (R_B removed), and Table 2-1 to find δ_B, the deflection at B due to the redundant R_B.

$$\Delta_B = \frac{w(2L)^4}{EI} 0.013021$$

$$\delta_B = \frac{R_B(2L)^3}{EI} 0.020833$$

By Eq. (4-2)
$$R_B = \frac{5}{4}wL$$

$$R_A = R_C = \frac{3}{8}wL$$

$$M_{BA} = -\frac{wL^2}{8}$$

$$M_{BC} = +\frac{wL^2}{8}$$ (ans.)

The deflection at D due to a uniformly distributed load over the whole span with $C'' = 0.013021$ (from Table 2-3) and $C_{BA} = 0.0625$ (from Table 3-2) is, by Eq. (3-39),

$$\Delta_D = \frac{L^3}{EI}\left(0.013021wL - 0.0625 \times \frac{1}{8}wL\right)$$

$$= \frac{L^3}{EI} 0.005209 \quad \text{in downward} \quad (ans.)$$

4 MOMENT-DISTRIBUTION METHOD

The moment-distribution method is widely used for analyzing statically indeterminate structures by distributing the fixed-end moments of beams and rigid frames.[2] It consists of solving simultaneous equations of the slope-deflection method by successive approximations. The degree of accuracy depends on the number of distributions made.

There are four factors required in the application of the moment-distribution method: (1) fixed-end moments, (2) rotational stiffness, (3) lateral stiffness, and (4) carryover factor. These four factors vary with the cross section of the members. When the member has a constant cross section, the application of the moment-distribution method is simple because all four factors are easy to find from simple formulas. But when the member has a variable cross section, these four factors need to be derived individually for each member. Let us illustrate these four factors for the member that has a constant cross section.

Fixed-end moments have been discussed in Chapter 3 and their formulas listed in Table 3-1.

The *rotational stiffness* is used for computing the moment-distribution factor. It is proportional to I/L. Assume that a member is allowed to rotate at one end while the other end is fixed. The stiffness factor is

$$K = 4E\frac{I}{L} \tag{4-4}$$

If the other end is simply supported or hinged instead of being fixed, the stiffness factor becomes

$$K = 3E\frac{I}{L} \tag{4-5}$$

This value of K is three-fourths of the stiffness of the same member having a fixed end. The values 4 or 3 are called *stiffness coefficients* and are denoted by k. In comparing the stiffness of various members, it may be seen that if these are uniform, but have varying cross sections, the relative stiffnesses can be represented by I/L. If the cross sections of the members are constant, then the relative stiffnesses are simply inversely proportional to the span length L.

Lateral stiffness can be defined as the moment development corresponding to a unit deformation at one end of the member. Assume, for example, that a beam of span length L and moment of inertia I has both ends fixed

---

[2] The moment-distribution method was originally presented by Prof. Hardy Cross in an article in *Trans. ASCE*, vol. 96, 1932.

against rotation. If the beam is displaced a distance Δ at one end, the fixed-end moment at the other end will be

$$M = 6E\Delta\frac{I}{L^2} \tag{4-6}$$

If one end is simply supported, then the fixed-end moment at the other end will be

$$M = 3E\Delta\frac{I}{L^2} \tag{4-7}$$

This is half the stiffness of the member having both ends fixed. When one end of a member is simply supported and the other end is fixed, the ratio of the moment at the fixed end to the moment producing the rotation at the simply supported end is called the *carryover factor*. The value is ½ when the member has a constant cross section. The relation between stiffness coefficient k and carryover factor C is

$$K = k\frac{EI}{L} \tag{4-8}$$

$$M = \frac{\Delta}{L}K(1 + C) \tag{4-9}$$

$$M = k\Delta\frac{EI}{L^2}(1 + C) \tag{4-10}$$

The moment distribution begins with these four factors and ends with the final moments. A check should be made to ensure that the correct answer has been obtained.

Example 4-5

Determine the support moments at B, from Example 4-3, using the moment-distribution method.

Solution

From Table 3-1, the fixed-end moments are given by

$$M_{AB} = \frac{PL}{8} \qquad M_{BA} = \frac{PL}{8}$$

$$M_{BC} = \frac{PL}{8} \qquad M_{CB} = \frac{PL}{8}$$

The moment-distribution information for Example 4-3 is given in the accompanying table.

Joint	A	B		C
Member	AB	BA	BC	CB
I/L		I/L	I/L	
Relative stiffness		1	1	
Moment-distribution factor		0.5	0.5	
Fixed-end moment	+0.125	−0.125	+0.125	−0.125
Balance	−0.125	—	—	+0.125
Carryover		−0.0625	+0.0625	
Balance		—	—	
Total*	0	−0.1875	+0.1875	0

*Support moments = total × PL.

From the table the support moments may be seen to be

$$M_{BA} = -\frac{3}{16}PL \qquad M_{BC} = +\frac{3}{16}PL \qquad (ans.)$$

These are the same answers as obtained in Example 4-3.

Example 4-6

Determine the support moments at B, from Example 4-4, using the moment-distribution method.

Solution

From Table 3-1, the fixed-end moments are given by

$$M_{AB} = \frac{1}{12}wL^2 \qquad M_{BA} = \frac{1}{12}wL^2$$

$$M_{BC} = \frac{1}{12}wL^2 \qquad M_{CB} = \frac{1}{12}wL^2$$

The moment-distribution information for Example 4-4 is given in the accompanying table.

Joint	A	B		C
Member	AB	BA	BC	CB
I/L		I/L	I/L	
Relative stiffness		1	1	
Moment-distribution factor		0.5	0.5	
Fixed-end moment	+0.083333	−0.083333	+0.083333	−0.083333
Balance	−0.083333	—	—	+0.083333
Carryover		−0.041667	+0.041667	
Balance		—	—	
Total*	0	−0.125	+0.125	0

*Support moments = total × wL^2.

From the table the support moments may be seen to be

$$M_{BA} = -\frac{1}{8}wL^2 \qquad M_{BC} = +\frac{1}{8}wL^2 \qquad (ans.)$$

These are the same answers as obtained in Example 4-4.

The moment distribution in these cases is very simple. We only need to carry over once. In most cases, the carryover has to be done several times to achieve an accurate result.

4-5 INFLUENCE LINES FOR DEFLECTION OF CONTINUOUS BEAMS

The foregoing discussion shows how we apply the support moments in the solution of the deflection of a continuous beam. In practice, it is more convenient to use influence line tables to find the coefficients of the deflection of a continuous beam, rather than compute the support moments, when the cross section of the beam is constant. The influence line coefficients from Tables 4-1 to 4-8, at the end of the chapter, are provided for this purpose.

The deflection found in these tables is the deflection at the center of the span, which is regarded as the maximum deflection of the span. The maximum deflection of a span for a continuous beam usually occurs between points 0.4 and 0.5, close to the center.

For example, assume that a continuous beam with two equal spans is loaded with a concentrated load P at the center of one span. The deflection at the center of the span is $\Delta_{0.5} = PL^3/EI \times 0.014974$ in (from Table 4-1), and the maximum deflection at the center of the span is $\Delta_{max} = PL^3/EI \times 0.0150$ in (see Appendix A, Case 36), occurring at $0.480L$ from the exterior support. The error is less than 0.2 percent. For a uniformly distributed load over the whole span, the deflection at the center of the span is $\Delta_{0.5} = wL^4/EI \times 0.009115$ in (from Table 4-5), and the maximum deflection of the span is $\Delta_{max} = wL^4/EI \times 0.0092$ in (see Appendix A, Case 38), occurring at $0.472L$ from the exterior support. The error is less than 1 percent. Therefore, the deflection at the center of the span is a good approximation of its maximum deflection. In using influence lines to determine the maximum deflection of one span of a continuous beam, the following procedure should be used.

Influence Lines for Concentrated Loads

To compute the maximum deflection of a continuous beam due to a series of concentrated loads, place the heaviest load at the largest influence line ordinate and sum up the ordinates occurring under the other loads. However, this procedure is not always usable. When there is more than one heavy load, the maximum deflection should be determined by the graphic solution. Tables 4-1 to 4-4 provide the influence line coefficients for concentrated loads.

Influence Lines for Uniformly Distributed Loads

In the case of a uniformly distributed load, the area of that portion of the influence line over which the load is situated should be determined and then multiplied by the magnitude of the load per unit length.

For a uniformly distributed load over the whole span, the total influence line area for the span should be obtained and multiplied by the load per unit length. Tables 4-5 to 4-8 are provided for such calculations.

In computing the area of a portion of the influence line for a span, Simpson's rule may be employed.

$$A = \frac{x}{3}[y_0 + y_n + 4(y_1 + y_3 + y_5 + \cdots) + 2(y_2 + y_4 + y_6 + \cdots)] \quad (4\text{-}11)$$

where $y_0, y_1, y_2, \ldots, y_n$ are the ordinates of the influence line of the area, and x is the space of the ordinates, equally divided.

Example 4-7

Using the influence lines, determine the deflection $\Delta_{0.5}$ for (a), (b), and (c) of Fig. 4-6. (Use Table 4-2.)

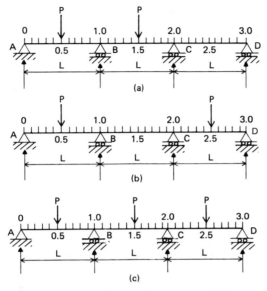

Fig. 4-6. Concentrated load on (a) spans 1 and 2; (b) spans 1 and 3; and (c) all spans.

Solution

(a) Three spans, concentrated load on spans 1 and 2

$$\Delta_{0.5} = \frac{PL^3}{EI}(0.014583 - 0.004688)$$

$$= \frac{PL^3}{EI} 0.009895 \quad \text{in downward} \quad (ans.)$$

(b) Three spans, concentrated load on spans 1 and 3

$$\Delta_{0.5} = \frac{PL^3}{EI}(0.014583 + 0.001563)$$

$$= \frac{PL^3}{EI} 0.016146 \quad \text{in downward} \quad (ans.)$$

(c) Three spans, concentrated load on all spans

$$\Delta_{0.5} = \frac{PL^3}{EI}(0.014583 - 0.004688 + 0.001563)$$

$$= \frac{PL^3}{EI} 0.011458 \quad \text{in downward} \quad (ans.)$$

The maximum deflection $\Delta_{0.5}$ occurs in the loading shown in Fig. 4-6(b).

Example 4-8

Using influence lines, determine the deflection $\Delta_{0.5}$ for (a), (b), and (c) of Fig. 4-7. (Use Table 4-6.)

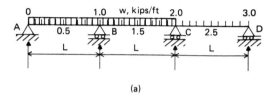

(a)

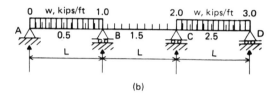

(b)

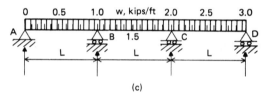

(c)

Fig. 4-7. Uniform load on (a) spans 1 and 2; (b) spans 1 and 3; and (c) all spans.

Solution

(a) Three spans, uniformly distributed load on spans 1 and 2

$$\Delta_{0.5} = \frac{wL^4}{EI}(0.008852 - 0.003125)$$

$$= \frac{wL^4}{EI} 0.005727 \quad \text{in downward} \quad (ans.)$$

(b) Three spans, uniformly distributed load on spans 1 and 3

$$\Delta_{0.5} = \frac{wL^4}{EI}(0.008852 + 0.001044)$$

$$= \frac{wL^4}{EI} 0.009896 \quad \text{in downward} \quad (ans.)$$

(c) Three spans, uniformly distributed load on all spans

$$\Delta_{0.5} = \frac{wL^4}{EI}(0.008852 - 0.003125 + 0.001044)$$

$$= \frac{wL^4}{EI} 0.006771 \quad \text{in downward} \quad (ans.)$$

The maximum deflection $\Delta_{0.5}$ occurs in the loading shown in Fig. 4-7(b).

Example 4-9

Using influence lines, determine the deflection $\Delta_{0.5}$ for (a), (b), and (c) of Fig. 4-8. (Use Table 4-3.)

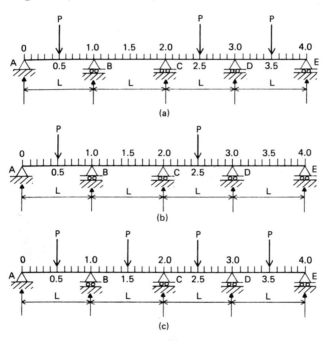

Fig. 4-8. Concentrated load on (a) spans 1, 3, and 4; (b) spans 1 and 3; and (c) all spans.

Solution

(a) Four spans, concentrated load on spans 1, 3, and 4

$$\Delta_{0.5} = \frac{PL^3}{EI}(0.014558 + 0.001256 - 0.000419)$$

$$= \frac{PL^3}{EI} 0.015395 \quad \text{in downward} \quad (ans.)$$

(b) Four spans, concentrated load on spans 1 and 3

$$\Delta_{0.5} = \frac{PL^3}{EI}(0.014558 + 0.001256)$$

$$= \frac{PL^3}{EI} 0.015814 \quad \text{in downward} \quad (ans.)$$

(c) Four spans, concentrated load on all spans

$$\Delta_{0.5} = \frac{PL^3}{EI}(0.014558 - 0.004606 + 0.001256 - 0.00419)$$

$$= \frac{PL^3}{EI} 0.010789 \quad \text{in downward} \quad (ans.)$$

The maximum deflection $\Delta_{0.5}$ occurs in the loading shown in Fig. 4-8(b).

Example 4-10

Using influence lines, determine the deflection $\Delta_{0.5}$ for (a), (b), and (c) of Fig. 4-9. (Use Table 4-7.)

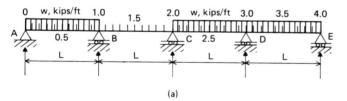

(a)

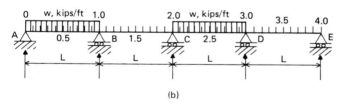

(b)

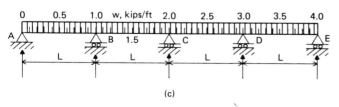

(c)

Fig. 4-9. Uniform load on (a) spans 1, 3, and 4; (b) spans 1 and 3; and (c) all spans.

Solution

(a) Four spans, uniformly distributed load on spans 1, 3, and 4

$$\Delta_{0.5} = \frac{wL^4}{EI}(0.008836 + 0.000838 - 0.000281)$$

$$= \frac{wL^4}{EI} 0.009393 \quad \text{in downward} \quad (ans.)$$

(b) Four spans, uniformly distributed load on spans 1 and 3

$$\Delta_{0.5} = \frac{wL^4}{EI}(0.008836 + 0.000838)$$

$$= \frac{wL^4}{EI} 0.009674 \quad \text{in downward} \quad (ans.)$$

(c) Four spans, uniformly distributed load on all spans

$$\Delta_{0.5} = \frac{wL^4}{EI}(0.008836 - 0.003069 + 0.000838 - 0.000281)$$

$$= \frac{wL^4}{EI} 0.006324 \quad \text{in downward} \quad (ans.)$$

The maximum deflection $\Delta_{0.5}$ occurs in the loading shown in Fig. 4-9(b).

From the above examples, it is seen that the maximum deflection for three equal spans, as in Examples 4-7 and 4-8, occurs in the (b) loading, in which spans 1 and 3 are loaded. Similarly, for four equal spans, as in Examples 4-9 and 4-10, the maximum deflection also occurs in the (b) loading. These examples depict a continuous beam of equal spans. When the spans are not equal, maximum deflection will depend on the loading condition.

If the applied loads are equal and positioned on spans 1 and 3, the maximum deflection will be on span 1, unless span 2 is longer than span 1 (in which case maximum deflection may occur on span 2). Similarly, when the spans and loads are equal, spans 2 and 4 must be loaded in order to obtain a maximum deflection on span 2.

If the applied loads are not equal, the deflection for the loading of each span will have to be calculated individually.

TABLE 4-1 INFLUENCE LINES FOR CONCENTRATED LOAD—TWO SPANS

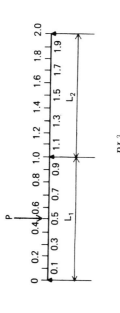

$$\Delta_x = \frac{PL_1^3}{EI} C$$

Position of unit load	Coefficient C									
	At point 0.5	At point 1.5	At point 0.5	At point 1.5	At point 0.5	At point 1.5	At point 0.5	At point 1.5
	$L_1:L_2 = 1:1$		$L_1:L_2 = 1:1.1$		$L_1:L_2 = 1:1.2$		$L_1:L_2 = 1:1.3$	
0	0	0	0	0	0	0	0	0
0.1	0.004619	−0.001550	0.004694	−0.001785	0.004761	−0.002025	0.004823	−0.002275
0.2	0.008833	−0.003002	0.008977	−0.003458	0.009108	−0.003929	0.009227	−0.004410
0.3	0.012232	−0.004269	0.012437	−0.004919	0.012625	−0.005582	0.012794	−0.006270
0.4	0.014417	−0.005253	0.014667	−0.006051	0.014896	−0.006876	0.015104	−0.007717
0.5	0.014974	−0.005860	0.015252	−0.006753	0.015508	−0.007669	0.015739	−0.008613
0.6	0.013667	−0.006003	0.013954	−0.006915	0.014214	−0.007857	0.014451	−0.008820
0.7	0.010919	−0.005581	0.011187	−0.006431	0.011431	−0.007301	0.011651	−0.008200
0.8	0.007333	−0.004502	0.007547	−0.005188	0.007743	−0.005895	0.007921	−0.006614
0.9	0.003494	−0.002675	0.003623	−0.003081	0.003739	−0.003501	0.003844	−0.003929

Span 1

Note: Minus sign means upward deflection.

TABLE 4-1 CONTINUED

Coefficient C

Position of unit load	At point 0.5	At point 1.5	At point 0.5	At point 1.5	At point 0.5	At point 1.5	At point 0.5	At point 1.5
	$L_1:L_2 = 1:1$		$L_1:L_2 = 1:1.1$		$L_1:L_2 = 1:1.2$		$L_1:L_2 = 1:1.3$	
1.0	0	0	0	0	0	0	0	0
1.1	−0.002675	0.003494	−0.003081	0.004484	−0.003501	0.005621	−0.003929	0.006916
1.2	−0.004502	0.007333	−0.005188	0.009477	−0.005895	0.011966	−0.006614	0.014822
1.3	−0.005581	0.010919	−0.006431	0.014187	−0.007301	0.018000	−0.008200	0.022395
1.4	−0.006003	0.013667	−0.006915	0.017813	−0.007857	0.022677	−0.008820	0.028304
1.5	−0.005860	0.014974	−0.006753	0.019561	−0.007669	0.024956	−0.008613	0.031215
1.6	−0.005253	0.014417	−0.006051	0.018859	−0.006876	0.024090	−0.007717	0.030174
1.7	−0.004269	0.012232	−0.004919	0.016016	−0.005582	0.020475	−0.006270	0.025657
1.8	−0.003002	0.008833	−0.003458	0.011568	−0.003926	0.014795	−0.004410	0.018550
1.9	−0.001550	0.004619	−0.001785	0.006053	−0.002025	0.007742	−0.002275	0.009706
2.0	0	0	0	0	0	0	0	0
	$L_1:L_2 = 1:1.4$		$L_1:L_2 = 1:1.5$		$L_1:L_2 = 1:1.6$		$L_1:L_2 = 1:1.7$	
0	0	0	0	0	0	0	0	0
0.1	0.004881	−0.002526	0.004929	−0.002788	0.004979	−0.003046	0.005023	−0.003314
0.2	0.009334	−0.004900	0.009433	−0.005402	0.009527	−0.005907	0.009609	−0.006430
0.3	0.012946	−0.006970	0.013087	−0.007681	0.013219	−0.008400	0.013338	−0.009140
0.4	0.015293	−0.008575	0.015467	−0.009453	0.015629	−0.010338	0.015779	−0.011248
0.5	0.015952	−0.009569	0.016145	−0.010550	0.016327	−0.011538	0.016489	−0.012553
0.6	0.014669	−0.009805	0.014867	−0.010803	0.015054	−0.011819	0.015223	−0.012855
0.7	0.011853	−0.009114	0.012037	−0.010044	0.012210	−0.010992	0.012369	−0.011951
0.8	0.008084	−0.007350	0.008233	−0.008102	0.008372	−0.008864	0.008502	−0.009641
0.9	0.003942	−0.004363	0.004029	−0.004813	0.004112	−0.005264	0.004186	−0.005726

Span 2 (rows 1.0–2.0); Span 1 (rows 0–0.9)

98

	$L_1:L_2 = 1:1.8$		$L_1:L_2 = 1:1.9$		$L_1:L_2 = 1:2$		$L_1:L_2 = 1:2.5$	
Span 2								
1.1	−0.004363	0.008371	−0.004813	0.009990	−0.005264	0.011792	−0.005726	0.013771
1.2	−0.007350	0.018066	−0.008102	0.021711	−0.008864	0.025784	−0.009641	0.030302
1.3	−0.009114	0.027420	−0.010044	0.033090	−0.010992	0.039466	−0.011951	0.046554
1.4	−0.009805	0.034758	−0.010803	0.042076	−0.011819	0.050311	−0.012855	0.059506
1.5	−0.009569	0.038411	−0.010550	0.046573	−0.011538	0.055796	−0.012553	0.066102
1.6	−0.008575	0.037162	−0.009453	0.045113	−0.010338	0.054092	−0.011248	0.064148
1.7	−0.006970	0.031622	−0.007681	0.038407	−0.008400	0.046082	−0.009140	0.054676
1.8	−0.004900	0.022868	−0.005402	0.027786	−0.005907	0.033348	−0.006430	0.039575
1.9	−0.002526	0.011973	−0.002788	0.014547	−0.003046	0.017488	−0.003314	0.020728
2.0	0	0	0	0	0	0	0	0
Span 1								
0	0	0	0	0	0	0	0	0
0.1	0.005061	−0.003584	0.005099	−0.003858	0.005136	−0.004125	0.005286	−0.005528
0.2	0.009689	−0.006947	0.009764	−0.007474	0.009833	−0.008000	0.010120	−0.010720
0.3	0.013456	−0.009877	0.013556	−0.010627	0.013656	−0.011375	0.014062	−0.015243
0.4	0.015917	−0.012156	0.016048	−0.013080	0.016167	−0.014000	0.016667	−0.018760
0.5	0.016645	−0.013568	0.016789	−0.014598	0.016927	−0.015625	0.017483	−0.020938
0.6	0.015379	−0.013893	0.015529	−0.014940	0.015667	−0.016000	0.016237	−0.021445
0.7	0.012513	−0.012920	0.012650	−0.013899	0.012781	−0.014875	0.013312	−0.019924
0.8	0.008620	−0.010420	0.008727	−0.011214	0.008833	−0.012000	0.009264	−0.016080
0.9	0.004261	−0.006187	0.004323	−0.006657	0.004386	−0.007125	0.004642	−0.009548
1.0	0	0	0	0	0	0	0	0
Span 2								
1.1	−0.006187	0.015939	−0.006657	0.018292	−0.007125	0.020836	−0.009548	0.036723
1.2	−0.010420	0.035273	−0.011214	0.040719	−0.012000	0.046664	−0.016080	0.084461
1.3	−0.012920	0.054395	−0.013899	0.063041	−0.014875	0.072500	−0.019924	0.133323
1.4	−0.013893	0.069702	−0.014940	0.080972	−0.016000	0.093336	−0.021445	0.173377
1.5	−0.013568	0.077555	−0.014598	0.090233	−0.015625	0.104164	−0.020938	0.194735
1.6	−0.012156	0.075332	−0.013080	0.087718	−0.014000	0.101336	−0.018760	0.190117
1.7	−0.009877	0.064238	−0.010627	0.074840	−0.011375	0.086500	−0.015243	0.162604
1.8	−0.006947	0.046512	−0.007474	0.054201	−0.008000	0.062664	−0.010720	0.117983
1.9	−0.003584	0.024365	−0.003858	0.028400	−0.004125	0.032836	−0.005528	0.061833
2.0	0	0	0	0	0	0	0	0

Note: Minus sign means upward deflection.

TABLE 4-2 INFLUENCE LINES FOR CONCENTRATED LOAD—THREE SPANS

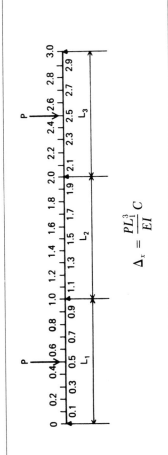

$$\Delta_x = \frac{PL_1^3}{EI} C$$

Position of unit load		Coefficient C					
		At point 0.5	At point 1.5	At point 0.5	At point 1.5	At point 0.5	At point 1.5
		$L_1:L_2:L_3 = 1:1:1$		$L_1:L_2:L_3 = 1:1.1:1$		$L_1:L_2:L_3 = 1:1.2:1$	
	0	0	0	0	0	0	0
Span 1	0.1	0.004517	−0.001238	0.004586	−0.001413	0.004648	−0.001592
	0.2	0.008633	−0.002400	0.008764	−0.002742	0.008888	−0.003082
	0.3	0.011950	−0.003413	0.012137	−0.003895	0.012311	−0.004385
	0.4	0.014067	−0.004200	0.014298	−0.004794	0.014510	−0.005399
	0.5	0.014583	−0.004688	0.014839	−0.005352	0.015077	−0.006028
	0.6	0.013267	−0.004800	0.013529	−0.005480	0.013773	−0.006172
	0.7	0.010550	−0.004463	0.010794	−0.005096	0.011020	−0.005739
	0.8	0.007033	−0.003600	0.007232	−0.004108	0.007413	−0.004627
	0.9	0.003317	−0.002138	0.003435	−0.002438	0.003542	−0.002746

	$L_1:L_2:L_3 = 1:1.3:1$		$L_1:L_2:L_3 = 1:1.4:1$		$L_1:L_2:L_3 = 1:1.5:1$	
Span 2						
1.1	−0.002438	0.002791	−0.002806	0.003544	−0.003183	0.004411
1.2	−0.004000	0.005833	−0.004596	0.007463	−0.005208	0.009342
1.3	−0.004813	0.008624	−0.005521	0.011087	−0.006245	0.013934
1.4	−0.005000	0.010667	−0.005725	0.013746	−0.006464	0.017322
1.5	−0.004688	0.011457	−0.005352	0.014781	−0.006028	0.018647
1.6	−0.004000	0.010667	−0.004552	0.013746	−0.005114	0.017322
1.7	−0.003063	0.008624	−0.003471	0.011087	−0.003882	0.013934
1.8	−0.002000	0.005833	−0.002256	0.007463	−0.002508	0.009342
1.9	−0.000938	0.002791	−0.001050	0.003544	−0.001157	0.004411
2.0	0	0	0	0	0	0
Span 3						
2.1	0.000713	−0.002138	0.000717	−0.002438	0.000717	−0.002746
2.2	0.001200	−0.003600	0.001206	−0.004108	0.001206	−0.004627
2.3	0.001488	−0.004463	0.001494	−0.005096	0.001494	−0.005739
2.4	0.001600	−0.004800	0.001606	−0.005480	0.001606	−0.006172
2.5	0.001563	−0.004688	0.001569	−0.005352	0.001569	−0.006028
2.6	0.001400	−0.004200	0.001406	−0.004794	0.001406	−0.005399
2.7	0.001138	−0.003413	0.001143	−0.003895	0.001143	−0.004385
2.8	0.000800	−0.002400	0.000805	−0.002742	0.000805	−0.003082
2.9	0.000413	−0.001238	0.000414	−0.001413	0.000414	−0.001592
3.0	0	0	0	0	0	0

	$L_1:L_2:L_3 = 1:1.3:1$		$L_1:L_2:L_3 = 1:1.4:1$		$L_1:L_2:L_3 = 1:1.5:1$	
0	0	0	0	0	0	0
Span 1						
0.1	0.004705	−0.001774	0.004760	−0.001952	0.004806	−0.002147
0.2	0.008997	−0.003440	0.009101	−0.003790	0.009196	−0.004154
0.3	0.012469	−0.004889	0.012613	−0.005400	0.012751	−0.005907
0.4	0.014705	−0.006018	0.014886	−0.006641	0.015051	−0.007270
0.5	0.015295	−0.006715	0.015495	−0.007413	0.015683	−0.008114
0.6	0.013998	−0.006875	0.014203	−0.007587	0.014393	−0.008309
0.7	0.011230	−0.006390	0.011419	−0.007057	0.011596	−0.007729
0.8	0.007582	−0.005155	0.007734	−0.005695	0.007878	−0.006230
0.9	0.003642	−0.003062	0.003735	−0.003374	0.003818	−0.003699

Note: Minus sign means upward deflection.

TABLE 4-2 CONTINUED

Position of unit load	Coefficient C					
	$L_1:L_2:L_3 = 1:1.3:1$		$L_1:L_2:L_3 = 1:1.4:1$		$L_1:L_2:L_3 = 1:1.5:1$	
	At point 0.5	At point 1.5	At point 0.5	At point 1.5	At point 0.5	At point 1.5
1.0	0	0	0	0	0	0
1.1	−0.003570	0.005383	−0.003964	0.006467	−0.004368	0.007669
1.2	−0.005833	0.011475	−0.006471	0.013880	−0.007120	0.016575
1.3	−0.006984	0.017195	−0.007738	0.020870	−0.008501	0.025028
1.4	−0.007215	0.021428	−0.007982	0.026072	−0.008752	0.031331
1.5	−0.006715	0.023082	−0.007413	0.028108	−0.008114	0.033805
1.6	−0.005678	0.021428	−0.006249	0.026072	−0.006825	0.031331
1.7	−0.004297	0.017195	−0.004713	0.020870	−0.005127	0.025028
1.8	−0.002763	0.011475	−0.003013	0.013880	−0.003264	0.016575
1.9	−0.001265	0.005383	−0.001370	0.006467	−0.001475	0.007669
2.0	0	0	0	0	0	0
2.1	0.000713	−0.003062	0.000712	−0.003974	0.000705	−0.003699
2.2	0.001200	−0.005155	0.001195	−0.005695	0.001187	−0.006230
2.3	0.001488	−0.006390	0.001482	−0.007057	0.001470	−0.007729
2.4	0.001600	−0.006675	0.001594	−0.007587	0.001581	−0.008309
2.5	0.001563	−0.006715	0.001556	−0.007413	0.001544	−0.008114
2.6	0.001400	−0.006018	0.001394	−0.006641	0.001386	−0.007270
2.7	0.001138	−0.004889	0.001132	−0.005400	0.001125	−0.005907
2.8	0.000800	−0.003440	0.000799	−0.003790	0.000792	−0.004154
2.9	0.000413	−0.001774	0.000412	−0.001952	0.000407	−0.002147
3.0	0	0	0	0	0	0

Span 2: rows 1.1–1.9
Span 3: rows 2.1–2.9

		$L_1:L_2:L_3 = 1:1.6:1$		$L_1:L_2:L_3 = 1:1.7:1$		$L_1:L_2:L_3 = 1:1.8:1$	
	0	0	0	0	0	0	0
Span 1	0.1	0.004853	−0.002325	0.004894	−0.002524	0.004935	−0.002711
	0.2	0.009283	−0.004523	0.009365	−0.004889	0.009444	−0.005250
	0.3	0.012875	−0.006429	0.012993	−0.006940	0.013101	−0.007472
	0.4	0.015204	−0.007906	0.015349	−0.008554	0.015485	−0.009194
	0.5	0.015852	−0.008829	0.016014	−0.009538	0.016164	−0.010265
	0.6	0.014567	−0.009039	0.014735	−0.009760	0.014886	−0.010509
	0.7	0.011760	−0.008401	0.011913	−0.009083	0.012056	−0.009766
	0.8	0.008008	−0.006781	0.008133	−0.007329	0.008246	−0.007890
	0.9	0.003897	−0.004020	0.003969	−0.004351	0.004037	−0.004679
	1.0	0	0	0	0	0	0
Span 2	1.1	−0.004778	0.008999	−0.005193	0.010450	−0.005614	0.012028
	1.2	−0.007784	0.019553	−0.008451	0.022848	−0.009127	0.026447
	1.3	−0.009284	0.029631	−0.010063	0.034756	−0.010860	0.040374
	1.4	−0.009542	0.037180	−0.010329	0.043693	−0.011133	0.050854
	1.5	−0.008829	0.040148	−0.009538	0.047227	−0.010265	0.054996
	1.6	−0.007409	0.037180	−0.007986	0.043693	−0.008576	0.050854
	1.7	−0.005548	0.029631	−0.005961	0.034756	−0.006383	0.040374
	1.8	−0.003516	0.019553	−0.003761	0.022848	−0.004013	0.026447
	1.9	−0.001577	0.008999	−0.001676	0.010450	−0.001779	0.012028
	2.0	0	0	0	0	0	0
Span 3	2.1	0.000699	−0.004020	0.000694	−0.004351	0.000686	−0.004679
	2.2	0.001176	−0.006781	0.001165	−0.007329	0.001151	−0.007890
	2.3	0.001457	−0.008401	0.001445	−0.009083	0.001430	−0.009766
	2.4	0.001569	−0.009039	0.001556	−0.009760	0.001537	−0.010509
	2.5	0.001531	−0.008829	0.001519	−0.009538	0.001500	−0.010265
	2.6	0.001374	−0.007906	0.001359	−0.008554	0.001344	−0.009194
	2.7	0.001114	−0.006429	0.001106	−0.006940	0.001093	−0.007472
	2.8	0.000783	−0.004523	0.000777	−0.004889	0.000769	−0.005250
	2.9	0.000405	−0.002325	0.000401	−0.002524	0.000395	−0.002711
	3.0	0	0	0	0	0	0

Note: Minus sign means upward deflection.

TABLE 4-2 CONTINUED

Position of unit load	Coefficient C					
	At point 0.5	At point 1.5		At point 0.5	At point 1.5	
	$L_1:L_2:L_3 = 1:1.9:1$			$L_1:L_2:L_3 = 1:2:1$		
	0	0		0	0	
Span 1						
0.1	0.004973	−0.002895		0.005006	−0.003097	
0.2	0.009515	−0.005622		0.009583	−0.006000	
0.3	0.013206	−0.007993		0.013301	−0.008528	
0.4	0.015612	−0.009841		0.015730	−0.010500	
0.5	0.016308	−0.010988		0.016439	−0.011722	
0.6	0.015031	−0.011255		0.015168	−0.012000	
0.7	0.012190	−0.010465		0.012318	−0.011153	
0.8	0.008358	−0.008439		0.008459	−0.009000	
0.9	0.004104	−0.005010		0.004162	−0.005347	
1.0	0	0		0	0	
Span 2						
1.1	−0.006038	0.013738		−0.006470	0.015586	
1.2	−0.009810	0.030384		−0.010502	0.034664	
1.3	−0.011661	0.046531		−0.012471	0.053250	
1.4	−0.011937	0.058732		−0.012753	0.067336	
1.5	−0.010988	0.063564		−0.011722	0.072916	
1.6	−0.009162	0.058732		−0.009754	0.067336	
1.7	−0.006801	0.046531		−0.007223	0.053250	
1.8	−0.004257	0.030384		−0.004503	0.034664	
1.9	−0.001875	0.013738		−0.001971	0.015586	

2.0		0	0	0	0	0
2.1		0.000675	−0.005010	0.000668	−0.005347
2.2		0.001138	−0.008439	0.001124	−0.009000
2.3		0.001412	−0.010465	0.001393	−0.011153
2.4	Span 3	0.001518	−0.011255	0.001499	−0.012000
2.5		0.001481	−0.010988	0.001463	−0.011722
2.6		0.001329	−0.009841	0.001312	−0.010500
2.7		0.001080	−0.007993	0.001067	−0.008528
2.8		0.000761	−0.005622	0.000750	−0.006000
2.9		0.000393	−0.002895	0.000387	−0.003097
3.0		0	0	0	0

Note: Minus sign means upward deflection.

TABLE 4-3 INFLUENCE LINES FOR CONCENTRATED LOAD—FOUR SPANS

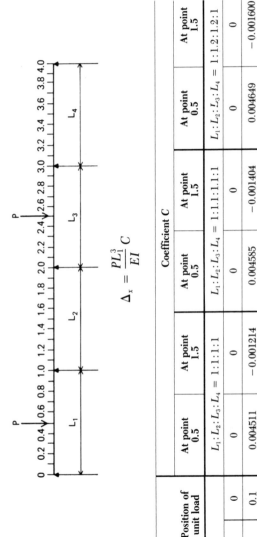

$$\Delta_x = \frac{PL_1^3}{EI} C$$

Position of unit load	Coefficient C			
	At point 0.5	At point 1.5	At point 0.5	At point 1.5
	$L_1:L_2:L_3:L_4 = 1:1:1:1$		$L_1:L_2:L_3:L_4 = 1:1:1:1$	
0	0	0	0	0
0.1	0.004511	−0.001214	0.004585	−0.001404
0.2	0.008620	−0.002358	0.008760	−0.002723
0.3	0.011932	−0.003353	0.012132	−0.003872
0.4	0.014044	−0.004127	0.014292	−0.004766
0.5	0.014558	−0.004606	0.014833	−0.005318
0.6	0.013239	−0.004717	0.013523	−0.005446
0.7	0.010526	−0.004385	0.010788	−0.005062
0.8	0.007014	−0.003537	0.007226	−0.004085
0.9	0.003305	−0.002100	0.003430	−0.002426

Position of unit load	At point 0.5	At point 1.5
	$L_1:L_2:L_3:L_4 = 1:1.2:1.2:1$	
0	0	0
0.1	0.004649	−0.001600
0.2	0.008889	−0.003103
0.3	0.012317	−0.004405
0.4	0.014516	−0.005427
0.5	0.015083	−0.006057
0.6	0.013779	−0.006202
0.7	0.011029	−0.005763
0.8	0.007419	−0.004652
0.9	0.003547	−0.002758

Span 1

Position									
1.0	0	0	0	0	0	0	0	0	0
1.1	−0.002419	0.002741	−0.002799	0.003522	−0.003189	0.004430
1.2	−0.003963	0.005724	−0.004581	0.007416	−0.005220	0.009386
1.3	−0.004756	0.008460	−0.005499	0.011012	−0.006264	0.014005
1.4	−0.004931	0.010448	−0.005694	0.013653	−0.006488	0.017417
1.5	−0.004606	0.011202	−0.005318	0.014668	−0.006057	0.018755
1.6	−0.003913	0.010396	−0.004517	0.013625	−0.005144	0.017439
1.7	−0.002975	0.008366	−0.003437	0.010972	−0.003913	0.014049
1.8	−0.001931	0.005615	−0.002225	0.007369	−0.002532	0.009434
1.9	−0.000894	0.002660	−0.001031	0.003489	−0.001174	0.004467
2.0	0	0	0	0	0	0
2.1	0.000650	−0.001961	0.000750	−0.002564	0.000850	−0.003275
2.2	0.001069	−0.003214	0.001231	−0.004199	0.001394	−0.005361
2.3	0.001288	−0.003869	0.001481	−0.005042	0.001675	−0.006425
2.4	0.001338	−0.004019	0.001531	−0.005233	0.001731	−0.006649
2.5	0.001256	−0.003769	0.001431	−0.004893	0.001613	−0.006201
2.6	0.001069	−0.003218	0.001219	−0.004161	0.001369	−0.005258
2.7	0.000819	−0.002462	0.000931	−0.003171	0.001038	−0.003995
2.8	0.000538	−0.001607	0.000606	−0.002060	0.000669	−0.002582
2.9	0.000250	−0.000755	0.000281	−0.000960	0.000313	−0.001190
3.0	0	0	0	0	0	0
3.1	−0.000193	0.000569	−0.000193	0.000651	−0.000193	0.000738
3.2	−0.000321	0.000969	−0.000324	0.001097	−0.000324	0.001233
3.3	−0.000400	0.001194	−0.000400	0.001369	−0.000400	0.001539
3.4	−0.000430	0.001282	−0.000430	0.001467	−0.000430	0.001656
3.5	−0.000419	0.001256	−0.000419	0.001437	−0.000419	0.001620
3.6	−0.000375	0.001125	−0.000375	0.001285	−0.000375	0.001449
3.7	−0.000306	0.000913	−0.000306	0.001043	−0.000306	0.001179
3.8	−0.000214	0.000643	−0.000214	0.000741	−0.000214	0.000828
3.9	−0.000112	0.000331	−0.000112	0.000378	−0.000112	0.000423
4.0	0	0	0	0	0	0

Span 2 : rows 1.0–2.0
Span 3 : rows 2.0–3.0
Span 4 : rows 3.0–4.0

Note: Minus sign means upward deflection.

TABLE 4-3 CONTINUED

Position of unit load	Coefficient C					
	At point 0.5	At point 1.5	At point 0.5	At point 1.5	At point 0.5	At point 1.5
	$L_1:L_2:L_3:L_4 = 1:1.3:1.3:1$		$L_1:L_2:L_3:L_4 = 1:1.4:1.4:1$		$L_1:L_2:L_3:L_4 = 1:1.5:1.5:1$	
0	0	0	0	0	0	0
Span 1						
0.1	0.004712	−0.001798	0.004768	−0.002006	0.004822	−0.002209
0.2	0.009010	−0.003486	0.009120	−0.003885	0.009221	−0.004289
0.3	0.012487	−0.004957	0.012643	−0.005524	0.012787	−0.006093
0.4	0.014726	−0.006105	0.014918	−0.006798	0.015098	−0.007500
0.5	0.015320	−0.006813	0.015533	−0.007586	0.015733	−0.008371
0.6	0.014023	−0.006974	0.014241	−0.007768	0.014444	−0.005577
0.7	0.011251	−0.006486	0.011455	−0.007219	0.011645	−0.007974
0.8	0.007597	−0.005236	0.007764	−0.005823	0.007915	−0.006436
0.9	0.003654	−0.003106	0.003749	−0.003464	0.003842	−0.003815
1.0	0	0	0	0	0	0
Span 2						
1.1	−0.003588	0.005459	−0.003998	0.006621	−0.004413	0.007928
1.2	−0.005874	0.011646	−0.006543	0.014221	−0.007224	0.017127
1.3	−0.007049	0.017462	−0.007849	0.021413	−0.008666	0.025892
1.4	−0.007296	0.021787	−0.008129	0.026789	−0.008970	0.032487
1.5	−0.006813	0.023504	−0.007586	0.028966	−0.008371	0.035183
1.6	−0.005783	0.021882	−0.006436	0.026987	−0.007103	0.032820
1.7	−0.004400	0.017631	−0.004893	0.021758	−0.005398	0.026467
1.8	−0.002846	0.011844	−0.003167	0.014615	−0.003491	0.017785
1.9	−0.001318	0.005607	−0.001463	0.006922	−0.001613	0.008419

	x	$L_1:L_2:L_3:L_4 = 1:1.6:1.6:1$		$L_1:L_2:L_3:L_4 = 1:1.7:1.7:1$		$L_1:L_2:L_3:L_4 = 1:1.8:1.8:1$	
Span 3	2.1	0.000956	−0.004099	0.001063	−0.005053	0.001169	−0.006135
	2.2	0.001563	−0.006704	0.001731	−0.008257	0.001906	−0.010004
	2.3	0.001869	−0.008025	0.002069	−0.009873	0.002275	−0.011952
	2.4	0.001931	−0.008288	0.002138	−0.010180	0.002344	−0.012297
	2.5	0.001800	−0.007711	0.001981	−0.009457	0.002169	−0.011404
	2.6	0.001519	−0.006523	0.001675	−0.007968	0.001825	−0.009593
	2.7	0.001150	−0.004932	0.001263	−0.006009	0.001375	−0.007205
	2.8	0.000738	−0.003175	0.000806	−0.003848	0.000875	−0.004587
	2.9	0.000338	−0.001455	0.000369	−0.001752	0.000394	−0.002071
	3.0	0	0	0	0	0	0
Span 4	3.1	−0.000193	0.000814	−0.000190	0.000906	−0.000188	0.000998
	3.2	−0.000324	0.001373	−0.000321	0.001531	−0.000318	0.001660
	3.3	−0.000400	0.001711	−0.000400	0.001887	−0.000394	0.002067
	3.4	−0.000430	0.001837	−0.000427	0.002034	−0.000424	0.002222
	3.5	−0.000419	0.001795	−0.000419	0.001984	−0.000413	0.002180
	3.6	−0.000375	0.001616	−0.000375	0.001776	−0.000370	0.001954
	3.7	−0.000306	0.001309	−0.000306	0.001434	−0.000300	0.001589
	3.8	−0.000214	0.000930	−0.000214	0.001016	−0.000212	0.001111
	3.9	−0.000112	0.000475	−0.000112	0.000514	−0.000111	0.000563
	4.0	0	0	0	0	0	0
Span 1	0	0	0	0	0	0	0
	0.1	0.004868	−0.002428	0.004916	−0.002635	0.004955	−0.002856
	0.2	0.009315	−0.004703	0.009403	−0.005117	0.009484	−0.005544
	0.3	0.012919	−0.006694	0.013046	−0.007268	0.013161	−0.007877
	0.4	0.015261	−0.008225	0.015417	−0.008937	0.015559	−0.009687
	0.5	0.015914	−0.009183	0.016089	−0.009979	0.016245	−0.010827
	0.6	0.014630	−0.009405	0.014807	−0.010225	0.014971	−0.011078
	0.7	0.011818	−0.008739	0.011983	−0.009505	0.012132	−0.010303
	0.8	0.008057	−0.007046	0.008190	−0.007666	0.008313	−0.008306
	0.9	0.003924	−0.004790	0.004004	−0.004549	0.004074	−0.004937

Note: Minus sign means upward deflection.

TABLE 4-3 CONTINUED

Position of unit load	Coefficient C					
	At point 0.5	At point 1.5	At point 0.5	At point 1.5	At point 0.5	At point 1.5
	$L_1:L_2:L_3:L_4 = 1:1.6:1.6:1$		$L_1:L_2:L_3:L_4 = 1:1.7:1.7:1$		$L_1:L_2:L_3:L_4 = 1:1.8:1.8:1$	
Span 2						
1.0	0	0	0	0	0	0
1.1	−0.004840	0.009367	−0.005267	0.010974	−0.005710	0.012695
1.2	−0.007926	0.020361	−0.008615	0.024016	−0.009348	0.027945
1.3	−0.009504	0.030922	−0.010334	0.036580	−0.011211	0.042749
1.4	−0.009839	0.038895	−0.010693	0.046136	−0.011602	0.054053
1.5	−0.009183	0.042196	−0.009979	0.050116	−0.010827	0.058824
1.6	−0.007790	0.039391	−0.008465	0.046824	−0.009184	0.055004
1.7	−0.005921	0.031780	−0.006432	0.037788	−0.006977	0.044412
1.8	−0.003828	0.021361	−0.004156	0.025400	−0.004511	0.029849
1.9	−0.001770	0.010106	−0.001918	0.012014	−0.002084	0.014117
Span 3						
2.0	0	0	0	0	0	0
2.1	0.001275	−0.007358	0.001388	−0.008740	0.001500	−0.010248
2.2	0.002081	−0.011982	0.002263	−0.014208	0.002438	−0.016663
2.3	0.002481	−0.014296	0.002694	−0.016932	0.002900	−0.019834
2.4	0.002550	−0.014694	0.002763	−0.017381	0.002975	−0.020325
2.5	0.002363	−0.013584	0.002550	−0.016057	0.002744	−0.018731
2.6	0.001981	−0.011403	0.002138	−0.013442	0.002288	−0.015659
2.7	0.001481	−0.008539	0.001594	−0.010041	0.001706	−0.011645
2.8	0.000938	−0.005416	0.001006	−0.006338	0.001069	−0.007322
2.9	0.000419	−0.002429	0.000450	−0.002821	0.000475	−0.003239

Note: Minus sign means upward deflection.

Span 4

x						
3.1	−0.000188	0.001072	−0.000187	0.001156	−0.000182	0.001256
3.2	−0.000315	0.001824	−0.000313	0.001951	−0.000307	0.002106
3.3	−0.000391	0.002256	−0.000388	0.002420	−0.000381	0.002613
3.4	−0.000421	0.002416	−0.000415	0.002619	−0.000412	0.002794
3.5	−0.000413	0.002352	−0.000406	0.002547	−0.000400	0.002734
3.6	−0.000370	0.002112	−0.000364	0.002294	−0.000357	0.002471
3.7	−0.000300	0.001712	−0.000295	0.001860	−0.000293	0.001984
3.8	−0.000212	0.001200	−0.000207	0.001301	−0.000206	0.001398
3.9	−0.000108	0.000624	−0.000107	0.000668	−0.000106	0.000729
4.0	0	0	0	0	0	0

$L_1:L_2:L_3:L_4 = 1:1.9:1.9:1$ \quad\quad $L_1:L_2:L_3:L_4 = 1:2:2:1$

Span 1

x				
0	0	0	0	0
0.1	0.004997	−0.003073	0.005031	−0.003300
0.2	0.009563	−0.005962	0.009632	−0.006400
0.3	0.013269	−0.008485	0.013370	−0.009100
0.4	0.015692	−0.010446	0.015816	−0.011200
0.5	0.016395	−0.011660	0.016533	−0.012500
0.6	0.015123	−0.011936	0.015266	−0.012800
0.7	0.012275	−0.011099	0.012406	−0.011900
0.8	0.008426	−0.008966	0.008532	−0.009600
0.9	0.004143	−0.005320	0.004205	−0.005700
1.0	0	0	0	0

Span 2

x				
1.1	−0.006153	0.014587	−0.006601	0.016636
1.2	−0.010071	0.032310	−0.010802	0.037064
1.3	−0.012078	0.049596	−0.012951	0.057100
1.4	−0.012498	0.062868	−0.013401	0.072536
1.5	−0.011660	0.068505	−0.012500	0.079164
1.6	−0.009891	0.064113	−0.010599	0.074136
1.7	−0.007515	0.051785	−0.008049	0.059900
1.8	−0.004854	0.034812	−0.005198	0.040264
1.9	−0.002241	0.016461	−0.002399	0.019036

TABLE 4-3 CONTINUED

Position of unit load		Coefficient C			
		At point 0.5	At point 1.5	At point 0.5	At point 1.5
		$L_1:L_2:L_3:L_4 = 1:1.9:1.9:1$		$L_1:L_2:L_3:L_4 = 1:2:2:1$	
Span 3	2.0	0	0	0	0
	2.1	0.001613	−0.011938	0.001725	−0.013800
	2.2	0.002619	−0.019389	0.002800	−0.022400
	2.3	0.003113	−0.023046	0.003325	−0.026600
	2.4	0.003181	−0.023603	0.003400	−0.027200
	2.5	0.002931	−0.021728	0.003125	−0.025000
	2.6	0.002444	−0.018104	0.002600	−0.020800
	2.7	0.001813	−0.013447	0.001925	−0.015400
	2.8	0.001138	−0.008405	0.001200	−0.009600
	2.9	0.000500	−0.003703	0.000525	−0.004200
Span 4	3.0	0	0	0	0
	3.1	−0.000181	0.001332	−0.000181	0.001425
	3.2	−0.000304	0.002233	−0.000301	0.002400
	3.3	−0.000375	0.002798	−0.000375	0.002975
	3.4	−0.000405	0.003000	−0.000402	0.003200
	3.5	−0.000394	0.002934	−0.000394	0.003125
	3.6	−0.000355	0.002617	−0.000352	0.002800
	3.7	−0.000288	0.002143	−0.000288	0.002275
	3.8	−0.000201	0.001512	−0.000201	0.001600
	3.9	−0.000105	0.000767	−0.000105	0.000825
	4.0	0	0	0	0

Note: Minus sign means upward deflection.

TABLE 4-4 INFLUENCE LINES FOR CONCENTRATED LOAD—FIVE OR MORE EQUAL SPANS

$$\Delta_x = \frac{PL_1^3}{EI} C$$

Position of unit load		Value of C, any inner-span midpoint*
	0	0
Span 1	0.1	−0.000707
	0.2	−0.001519
	0.3	−0.002344
	0.4	−0.003081
	0.5	−0.003631
	0.6	−0.003881
	0.7	−0.003744
	0.8	−0.003125
	0.9	−0.001906
	1.0	0
Span 2	1.1	0.002599
	1.2	0.005489
	1.3	0.008178
	1.4	0.010153
	1.5	0.010921
	1.6	0.010153
	1.7	0.008178
	1.8	0.005489
	1.9	0.002599
	2.0	0
Span 3	2.1	−0.001908
	2.2	−0.003127
	2.3	−0.003749
	2.4	−0.003885
	2.5	−0.003631
	2.6	−0.003083
	2.7	−0.002346
	2.8	−0.001520
	2.9	−0.000706
	3.0	0
Span 4	3.1	0.000512
	3.2	0.000838
	3.3	0.001006
	3.4	0.001044
	3.5	0.000969
	3.6	0.000825
	3.7	0.000626
	3.8	0.000406
	3.9	0.000188
	4.0	0

*The values of C for the midpoint of the end span do not differ significantly from the values for the end span in Table 4-3. Use the same values of C as for the beam with span ratios of $1:1:1:1$.
Note: Minus sign means upward deflection.

TABLE 4-5 INFLUENCE LINES FOR SPAN ENTIRELY COVERED BY UNIFORMLY DISTRIBUTED LOAD—TWO SPANS

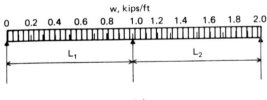

$$\Delta_x = \frac{wL_1^4}{EI} C$$

$L_1:L_2$	Deflection point	Total load, dead weight	Coefficient C Span 1 loaded	Span 2 loaded
1:1	0.5 1.5	0.005209 0.005209	0.009115 −0.003906	−0.003906 0.009115
1:1.1	0.5 1.5	0.004348 0.008570	0.009300 −0.004500	−0.004952 0.013070
1:1.2	0.5 1.5	0.003333 0.013050	0.009471 −0.005112	−0.006138 0.018162
1:1.3	0.5 1.5	0.002159 0.018831	0.009622 −0.005746	−0.007463 0.024577
1:1.4	0.5 1.5	0.000833 0.026134	0.009764 −0.006382	−0.008931 0.032516
1:1.5	0.5 1.5	−0.000654 0.035149	0.009896 −0.007031	−0.010550 0.042180
1:1.6	0.5 1.5	−0.002292 0.046134	0.010014 −0.007696	−0.012306 0.053830
1:1.7	0.5 1.5	−0.004092 0.059297	0.010127 −0.008363	−0.014219 0.067660
1:1.8	0.5 1.5	−0.006042 0.074926	0.010233 −0.009032	−0.016275 0.083958
1:1.9	0.5 1.5	−0.008154 0.093248	0.010327 −0.009724	−0.018481 0.102972
1:2	0.5 1.5	−0.010417 0.114583	0.010417 −0.010417	−0.020834 0.125000
1:2.5	0.5 1.5	−0.024089 0.276697	0.010790 −0.013945	−0.034879 0.290642

Note: Minus sign means upward deflection.

TABLE 4-6 INFLUENCE LINES FOR SPAN ENTIRELY COVERED BY UNIFORMLY DISTRIBUTED LOAD—THREE SPANS

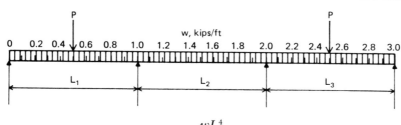

$$\Delta_x = \frac{wL_1^4}{EI} C$$

$L_1:L_2:L_3$	Deflection point	Total load, dead weight	Coefficient C Span 1 loaded	Span 2 loaded	Spans 1 and 3 loaded
1:1:1	0.5	0.006771	0.008852	−0.003125	0.009896
	1.5	0.000521	−0.003125	0.006771	−0.006250
	2.5	0.006771	0.001044	−0.003125	0.009896
1:1.1:1	0.5	0.006146	0.009027	−0.003925	0.010071
	1.5	0.002425	−0.003570	0.009565	−0.007140
	2.5	0.006146	0.001044	−0.003925	0.010071
1:1.2:1	0.5	0.005408	0.009183	−0.004819	0.010227
	1.5	0.005076	−0.004023	0.013122	−0.008046
	2.5	0.005408	0.001044	−0.004819	0.010227
1:1.3:1	0.5	0.004552	0.009327	−0.005819	0.010371
	1.5	0.008565	−0.004478	0.017521	−0.008956
	2.5	0.004552	0.001044	−0.005819	0.010371
1:1.4:1	0.5	0.003584	0.009462	−0.006916	0.010500
	1.5	0.013038	−0.004937	0.022912	−0.009874
	2.5	0.003584	0.001038	−0.006916	0.010500
1:1.5:1	0.5	0.002502	0.009587	−0.008116	0.010618
	1.5	0.018600	−0.005400	0.029400	−0.010800
	2.5	0.002502	0.001031	−0.008116	0.010618
1:1.6:1	0.5	0.001308	0.009702	−0.009413	0.010721
	1.5	0.025365	−0.005888	0.037141	−0.011776
	2.5	0.001308	0.001019	−0.009413	0.010721
1:1.7:1	0.5	0.000008	0.009808	−0.010813	0.010821
	1.5	0.033539	−0.006358	0.046255	−0.012716
	2.5	0.000008	0.001013	−0.010813	0.010821
1:1.8:1	0.5	−0.001405	0.009908	−0.012313	0.010908
	1.5	0.043813	−0.006845	0.056903	−0.013690
	2.5	−0.001405	0.001000	−0.012313	0.010908
1:1.9:1	0.5	−0.002929	0.010002	−0.013919	0.010990
	1.5	0.054529	−0.007333	0.069195	−0.014667
	2.5	−0.002929	0.000988	−0.013919	0.010990
1:2:1	0.5	−0.004560	0.010090	−0.015625	0.011065
	1.5	0.067683	−0.007825	0.083333	−0.015650
	2.5	−0.004560	0.000975	−0.015625	0.011065

Note: Minus sign means upward deflection.

TABLE 4-7 INFLUENCE LINES FOR SPAN ENTIRELY COVERED BY UNIFORMLY DISTRIBUTED LOAD—FOUR SPANS

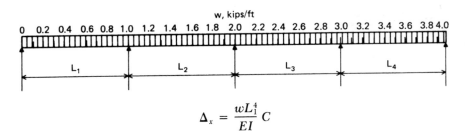

$$\Delta_x = \frac{wL_1^4}{EI} C$$

$L_1:L_2:L_3:L_4$	Deflection point	Coefficient C				
		Total load, dead weight	Span 1 loaded	Span 2 loaded	Spans 1 and 3 loaded	Spans 2 and 4 loaded
1:1:1:1	0.5	0.006324	0.008836	−0.003069	0.009674	0.007444
	1.5	0.001862	−0.003069	0.006606		
	2.5	0.001862	0.000838	−0.002513		
	3.5	0.006324	−0.000281	0.000838		
1:1.1:1.1:1	0.5	0.005887	0.009024	−0.003900	0.010068	0.010449
	1.5	0.003311	−0.003546	0.009487		
	2.5	0.003311	0.000962	−0.003592		
	3.5	0.005887	−0.000281	0.001044		
1:1.2:1.2:1	0.5	0.005359	0.009190	−0.004844	0.010484	0.014286
	1.5	0.005286	−0.004041	0.013200		
	2.5	0.005286	0.001086	−0.004959		
	3.5	0.005359	−0.000281	0.001294		
1:1.3:1.3:1	0.5	0.004721	0.009352	−0.005906	0.010908	0.019096
	1.5	0.007868	−0.004542	0.017886		
	2.5	0.007868	0.001210	−0.006686		
	3.5	0.004721	−0.000281	0.001556		
1:1.4:1.4:1	0.5	0.003984	0.009496	−0.007081	0.011346	0.025030
	1.5	0.011151	−0.005059	0.023696		
	2.5	0.011151	0.001334	−0.008820		
	3.5	0.003984	−0.000281	0.001850		
1:1.5:1.5:1	0.5	0.003140	0.009621	−0.008375	0.011790	0.032240
	1.5	0.015252	−0.005583	0.030782		
	2.5	0.015252	0.001458	−0.011405		
	3.5	0.003140	−0.000275	0.002169		
1:1.6:1.6:1	0.5	0.002197	0.009741	−0.009788	0.012260	0.040909
	1.5	0.020281	−0.006128	0.039327		
	2.5	0.020281	0.001582	−0.014500		
	3.5	0.002197	−0.000275	0.002519		
1:1.7:1.7:1	0.5	0.001152	0.009858	−0.011325	0.012746	0.051200
	1.5	0.026347	−0.006665	0.049494		
	2.5	0.026347	0.001706	−0.018188		
	3.5	0.001152	−0.000269	0.002888		

Note: Minus sign means upward deflection.

	Deflection point	Total load, dead weight	Coefficient C			
			Span 1 loaded	Span 2 loaded	Spans 1 and 3 loaded	Spans 2 and 4 loaded
$L_3:L_4$ 8:1	0.5	0.000003	0.009965	−0.012981	0.013253	
	1.5	0.033583	−0.007209	0.061466		0.063295
	2.5	0.033583	0.001829	−0.022503		
	3.5	0.000003	−0.000269	0.003288		
9:1	0.5	−0.001248	0.010065	−0.014763	0.013778	
	1.5	0.042125	−0.007762	0.075463		0.077415
	2.5	0.042125	0.001952	−0.027528		
	3.5	−0.001248	−0.000263	0.003713		
	0.5	−0.002604	0.010158	−0.016666	0.014325	
	1.5	0.052084	−0.008325	0.091667		0.093742
	2.5	0.052084	0.002075	−0.033333		
	3.5	−0.002604	−0.000263	0.004167		

Minus sign means upward deflection.

TABLE 4-8 INFLUENCE LINES FOR EQUAL SPANS ENTIRELY COVERED BY UNIFORMLY DISTRIBUTED LOAD—FIVE OR MORE SPANS

$$\Delta_x = \frac{wL_1^4}{EI} C$$

	Deflection point	Coefficient C	
		Total load, dead weight*	Any inner span loaded†
= 1	5	0.007815	−0.003300
	15	0.002609	0.006421
	25	0.002609	−0.002419
	35	0.002609	0.000881

*The coefficients are the same in all inner spans.
†The deflection in an end span is similar to that for the 1:1:1:1 beam.
Note: Minus sign means upward deflection.

5
Deflection of Cover-Plated Beams

EFFECT OF COVER PLATES

In practice, the beam cross section is not always constant. It varies due to the various loads and moments. The beam has to be reinforced by additional plates in some areas to give more strength so as to resist overloaded moments. For a beam whose cross section varies because of cover plates, the conjugate-beam method is very suitable in calculating the deflection of the beam. The procedure is the same as that illustrated for beams with constant cross section. The only difference is that the moment area on a conjugate beam has to be modified by the various moments of inertia at the various cross sections in order to obtain a modified moment diagram and then compute the deflection of the beam.

Example 5-1

A cantilever beam has cover plates at the support as shown in Fig. 5-1. A concentrated load of P kips is loaded at the end B. Find the deflection at that point by the conjugate-beam method. Assume that $b = pL$ (where $p = 0$ to 1) and $n = I_1/I_0$.

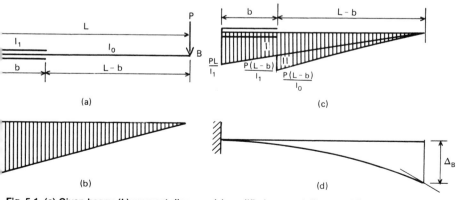

Fig. 5-1. (a) Given beam; (b) moment diagram; (c) modified moment diagram; (d) elastic curve.

Solution

$$\text{Area I} = \frac{PL^2}{2I_1} \quad \text{kip} \cdot \text{ft}^2$$

$$\text{Area II} = \frac{PL^2(1-p)^2}{2}\left(\frac{1}{I_0} - \frac{1}{I_1}\right) \quad \text{kip} \cdot \text{ft}^2$$

$$M'_B = \text{area I} \times \tfrac{2}{3}L + \text{area II} \times \tfrac{2}{3}(1-p)L$$

$$= \frac{PL^3}{3I_0}\left\{1 - [1-(1-p)^3]\left(\frac{n-1}{n}\right)\right\} \quad \text{kip} \cdot \text{ft}^3$$

$$\Delta_B = \frac{PL^3}{3EI_0}\left\{1 - [1-(1-p)^3]\left(\frac{n-1}{n}\right)\right\} \quad \text{in downward}$$

$$(ans.) \quad (5\text{-}1)$$

For an increase in the value of p, the amount of deflection is reduced. When $p = 1$, the cross section of the entire beam changes from I_0 to I_1. When $p = 0$, Eq. (5-1) becomes Eq. (1-4), the equation for a beam without cover plates and under a concentrated load.

Example 5-2

A cantilever beam has cover plates at the support as shown in Fig. 5-2. A uniformly distributed load covers the whole span. Find the deflection at the end of the beam by the conjugate-beam method. Assume that $b = pL$ (where $p = 0$ to 1) and $n = I_1/I_0$.

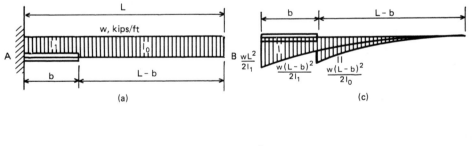

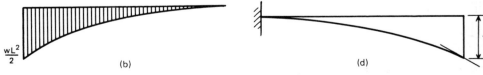

Fig. 5-2. (*a*) Given beam; (*b*) moment diagram; (*c*) modified moment diagram; (*d*) elastic curve.

Solution

$$\text{Area I} = \frac{wL^3}{6I_1} \quad \text{kip} \cdot \text{ft}^2$$

$$\text{Area II} = \frac{wL^3}{6}(1-p)^3\left(\frac{1}{I_0} - \frac{1}{I_1}\right) \quad \text{kip} \cdot \text{ft}^2$$

$$M'_B = \text{area I} \times \tfrac{3}{4}L + \text{area II} \times \tfrac{3}{4}(1-p)L$$

$$= \frac{wL^4}{8I_0}\left\{1 - [1-(1-p)^4]\left(\frac{n-1}{n}\right)\right\} \quad \text{kip} \cdot \text{ft}^3$$

$$\Delta_B = \frac{wL^4}{8EI_0}\left\{1 - [1-(1-p)^4]\left(\frac{n-1}{n}\right)\right\} \quad \text{in downward}$$

$$(ans.) \quad (5\text{-}2)$$

As in Eq. (5-1), when $p = 1$, the cross section of the entire beam changes from I_0 to I_1. When $p = 0$, Eq. (5-2) becomes Eq. (1-6), the equation for a beam without cover plates and under a uniformly distributed load.

COVER-PLATED BEAM WITH CONCENTRATED LOAD

Cover Plates at the Center of the Beam

Figure 5-3 shows a simple-support beam carrying a concentrated load of P kips at the center of the beam with cover plates at the center. In order to derive a general equation for the deflection Δ_D at the center of the beam, we assume that $a = mL$, $m = 0$ to 1, $b = \frac{1}{2}(L - a)$, and $n = I_1/I_0$. From Fig. 5-3(d) we can see that

$$\text{Area I} = \frac{1}{16I_0} P(L-a)^2 \quad \text{kip} \cdot \text{ft}^2$$

$$\text{Area II} = \frac{1}{16I_1} P(L-a)a \quad \text{kip} \cdot \text{ft}^2$$

$$\text{Area III} = \frac{1}{16I_1} PaL \quad \text{kip} \cdot \text{ft}^2$$

$$R'_A = R'_B = \frac{P}{16I_0}(L-a)^2 + \frac{P}{16I_1}a(2L-a) \quad \text{kip} \cdot \text{ft}^2$$

$$M'_D = R'_A \frac{L}{2} - \text{area I}\left[\frac{a}{2} + \frac{1}{6}(L-a)\right] - \text{area II}\,\frac{2}{3}\frac{a}{2} - \text{area III}\,\frac{1}{3}\frac{a}{2}$$

$$= \frac{PL^3}{48}\left[\frac{1}{I_0} - (3m - 3m^2 + m^3)\left(\frac{1}{I_0} - \frac{1}{I_1}\right)\right] \quad \text{kip} \cdot \text{ft}^3$$

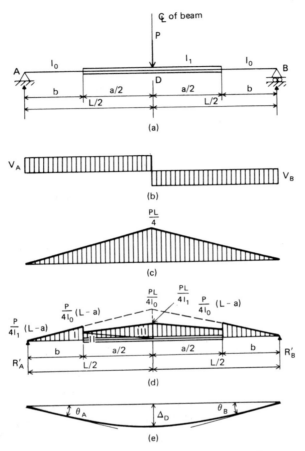

Fig. 5-3. (a) Given beam; (b) shear diagram; (c) moment diagram; (d) modified moment diagram; (e) elastic curve.

$$\Delta_D = \frac{PL^3}{EI_0}\left[\frac{1}{48} - \frac{1}{48}(3m - 3m^2 + m^3)\left(\frac{n-1}{n}\right)\right]$$

$$= \frac{PL^3}{EI_0}\left[C - f_1(m)\left(\frac{n-1}{n}\right)\right] \quad \text{in downward} \quad (5\text{-}3)$$

where, for a concentrated load at the center of the beam,

$$C = \frac{1}{48}$$

and
$$f_1(m) = \frac{1}{48}(3m - 3m^2 + m^2)$$

Equation (5-3) gives the deflection at the center of a simple-support beam with the cover plates at the center. The values of C and $f_1(m)$ vary with the position of the load. The reduction in deflection at the center of the beam due to the cover plates at the center is

$$\frac{PL^3}{EI_0} f_1(m) \left(\frac{n-1}{n}\right)$$

Cover Plates at the End of the Beam

Figure 5-4 shows a simple-support beam carrying a concentrated load of P kips at the center with cover plates at the left end of the beam. In order to derive a general equation for the deflection Δ_D at the center of the beam, we assume that $b = pL$, $p = 0$ to 0.5, and $n = I_1/I_0$. From Fig. 5-4(d) we can see that

$$\text{Area I} = \frac{1}{16I_0} PL^2 \quad \text{kip} \cdot \text{ft}^2$$

$$\text{Area II} = \frac{1}{4} Pb^2 \left(\frac{1}{I_0} - \frac{1}{I_1}\right) \quad \text{kip} \cdot \text{ft}^2$$

$$R'_A = \frac{PL^2}{16I_0} - \frac{PL^2}{12} p^2(3-2p)\left(\frac{1}{I_0} - \frac{1}{I_1}\right) \quad \text{kip} \cdot \text{ft}^2$$

$$R'_B = \frac{PL^2}{16I_0} - \frac{PL^2}{6} p^3 \left(\frac{1}{I_0} - \frac{1}{I_1}\right) \quad \text{kip} \cdot \text{ft}^2$$

$$M'_D = R'_B \frac{L}{2} - \text{area I} \frac{L}{6}$$

$$= \frac{PL^3}{48} \left[\frac{1}{I_0} - 4p^3\left(\frac{1}{I_0} - \frac{1}{I_1}\right)\right] \quad \text{kip} \cdot \text{ft}^3$$

$$\Delta_D = \frac{PL^3}{EI_0} \left[\frac{1}{48} - \frac{1}{12} p^3 \left(\frac{n-1}{n}\right)\right]$$

$$= \frac{PL^3}{EI_0} \left[C - f_1(p)\left(\frac{n-1}{n}\right)\right] \quad \text{in downward} \quad (5\text{-}4)$$

where, for a concentrated load at the center of the beam,

$$C = \frac{1}{48}$$

and

$$f_1(p) = \frac{1}{12} p^3$$

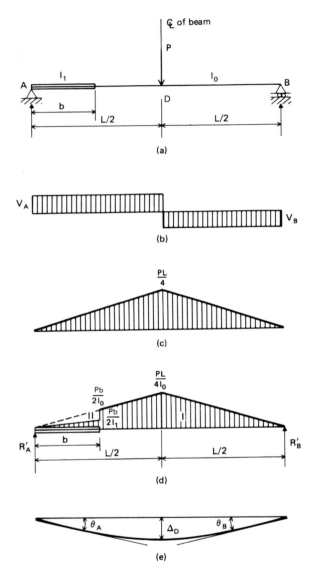

Fig. 5-4. (a) Given beam; (b) shear diagram; (c) moment diagram; (d) modified moment diagram; (e) elastic curve.

Equation (5-4) gives the deflection at the center of a simple-support beam with the cover plates at the left end. The values of C and $f_1(p)$ vary with the position of the load. The reduction in deflection at the center of the beam due to the cover plates at the left end of the beam is $(PL^3/EI_0)f_1(p)(n$

$- 1)/n$. If the cover plates are placed at the right end of the beam, use p' and b' to represent the right-end cover plates.

$$\Delta_D = \frac{PL^3}{EI_0}\left[C - f_1(p')\left(\frac{n-1}{n}\right)\right] \quad \text{in downward} \qquad (5\text{-}5)$$

A general equation for determining the deflection at the center of a simple-support, cover-plated beam, carrying a concentrated load on the beam, is derived by combining Eqs. (5-3), (5-4), and (5-5)

$$\Delta_D = \frac{PL^3}{EI_0}\left\{C - [f_1(m) + f_1(p) + f_1(p')]\left(\frac{n-1}{n}\right)\right\} \quad \text{in downward} \qquad (5\text{-}6)$$

where C = coefficient of deflection of a simple-support beam carrying a concentrated load (Table 2-1)

$f_1(m)$ = coefficient of deflection due to cover plates at the center of a simple-support beam carrying a concentrated load (Table 5-1)

$f_1(p)$ = coefficient of deflection due to cover plates at the left end of a simple-support beam carrying a concentrated load (Table 5-2)

$f_1(p')$ = coefficient of deflection due to cover plates at the right end of a simple-support beam carrying a concentrated load (Table 5-2)

Tables 5-1 and 5-2, at the end of the chapter, provide the coefficients for $f_1(m)$, $f_1(p)$ and $f_1(p')$. The deflection is at the center of the beam with a concentrated load applied at one of the one-tenth points along the beam. The deflections act upward to reduce the deflection downward from the load.

In Eq. (5-6) it is assumed that all cover plates have the same ratio for the moments of inertia. Otherwise, the value of $(n - 1)/n$ would be different for each part. The cover plates are also assumed to consist of one set of plates. If there is more than one set of cover plates, additional ratios of the moments of inertia should be used to obtain the additional reductions of deflection. This is illustrated in the following terms:

I_0 = moment of inertia of beam

I_k = moment of inertia of beam with a given number of cover plates, for example, I_1, I_2, I_3

n_k = ratio of the moments of inertia, for example, $n_1 = I_1/I_0$, $n_2 = I_2/I_1$, $n_3 = I_3/I_2$, $n_k = I_k/I_{k-1}$

$\dfrac{n_k - 1}{n_k!}$ = ratio of the moments of inertia of the various cover plate sets, for example, $(n_1 - 1)/n_1$, $(n_2 - 1)/n_1 n_2$, $(n_3 - 1)/n_1 n_2 n_3$

Example 5-6 illustrates the case where more than one set of cover plates is used.

Example 5-3

Find the deflection Δ_D for the concentrated load P applied at the center of a beam, as shown in Fig. 5-5, with $m = 0.5$, $n = I_1/I_0 = 2$, and $(n - 1)/n = 1/2$.

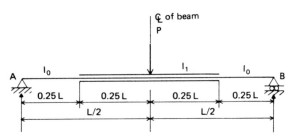

Fig. 5-5.

Solution

For the beam without cover plates, $C = 0.020833$ (from Table 2-1). For the beam with cover plates at the center, $f_1(m) = 0.018229$ (from Table 5-1). By Eq. (5-6),

$$\Delta_D = \frac{PL^3}{EI_0}\left(0.020833 - 0.018229 \times \frac{1}{2}\right)$$

$$= \frac{PL^3}{EI_0} 0.011719 \quad \text{in downward} \quad \text{(ans.)}$$

Example 5-4

Find the deflection Δ_D for the concentrated load P applied at the center of the beam, as shown in Fig. 5-6, with $p = 0.25$, $n = I_1/I_0 = 2$, and $(n - 1)/n = 1/2$.

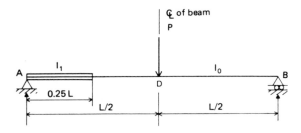

Fig. 5-6.

Solution

For the beam without cover plates, $C = 0.020833$ (from Table 2-1). For the beam with cover plates at the left end, $f_1(p) = 0.001302$ (from Table 5-2). By Eq. (5-6),

$$\Delta_D = \frac{PL^3}{EI_0}\left(0.020833 - 0.001302 \times \frac{1}{2}\right)$$

$$= \frac{PL^3}{EI_0} 0.020182 \quad \text{in downward} \quad (ans.)$$

COVER-PLATED BEAM WITH UNIFORMLY DISTRIBUTED LOAD

Cover Plates at the Center of the Beam

Figure 5-7 shows a simple-support beam carrying a uniformly distributed load over the whole span with cover plates at the center of the beam. In order to derive a general equation for the deflection Δ_D at the center of the beam, we assume that $a = mL$, $b = (L/2)(1 - m)$, $m = 0$ to 1, and $n = I_1/I_0$. From Fig. 5-7 we can see that

$$\text{Area I} = \frac{wL^3}{24I_1} \quad \text{kip} \cdot \text{ft}^2$$

$$\text{Area II} = \frac{wb^2}{12}(3L - 2b)\left(\frac{1}{I_0} - \frac{1}{I_1}\right) \quad \text{kip} \cdot \text{ft}^2$$

$$R'_A = R'_B = \frac{wL^3}{24I_1} + \frac{wb^2}{12}(3L - 2b)\left(\frac{1}{I_0} - \frac{1}{I_1}\right) \quad \text{kip} \cdot \text{ft}^2$$

$$M'_D = R'_A \frac{L}{2} - \text{area I}\left(\frac{L}{2} - \frac{5L}{16}\right) - \text{area II}\left[\frac{L}{2} - \frac{L(4L - 3b)}{2(3L - 2b)}\right]$$

$$= \frac{5wL^4}{384I_1} + \frac{wb^3}{24I_0}(4L - 3b)\left(\frac{1}{I_0} - \frac{1}{I_1}\right) \quad \text{kip} \cdot \text{ft}^3$$

$$\Delta_D = \frac{5wL^4}{384EI_0} - \frac{wL^4}{32EI_0}\left(m - \frac{m^2}{2} - \frac{m^3}{3} + \frac{m^4}{4}\right)\left(\frac{n-1}{n}\right)$$

$$= \frac{wL^4}{EI_0}\left[0.013021 - f_2(m)\left(\frac{n-1}{n}\right)\right] \quad \text{in downward} \quad (5\text{-}7)$$

where

$$f_2(m) = \frac{1}{32}\left(m - \frac{m^2}{2} - \frac{m^3}{3} + \frac{m^4}{4}\right)$$

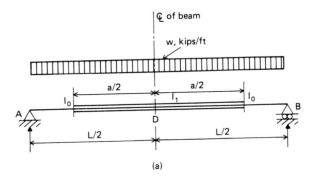

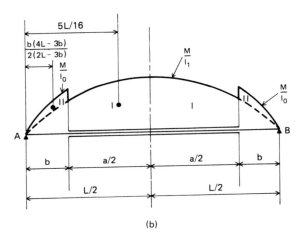

Fig. 5-7. (a) Given beam; (b) modified moment diagram.

The reduction in deflection at the center of the beam due to the cover plates at the center of the beam is $f_2(m)\,[(n-1)/n](wL^4/EI_0)$.

Cover Plates at the End of the Beam

Figure 5-8 shows a simple-support beam carrying a uniformly distributed load over the whole span with cover plates at the left end of the beam. In order to derive a general equation for the deflection at the center of the beam, we assume that $b = pL$, $p = 0$ to 0.5, and $n = I_1/I_0$. From Fig. 5-8 we can see that

$$\text{Area I} = \frac{wL^3}{24I_0} \quad \text{kip} \cdot \text{ft}^2$$

$$\text{Area II} = \frac{wb^2}{12}(3L - 2b)\left(\frac{1}{I_0} - \frac{1}{I_1}\right) \quad \text{kip} \cdot \text{ft}^2$$

$$R'_A = \frac{wL^3}{24I_0} - \frac{wb^3}{24}\left(6L - 8b + \frac{3b^2}{L}\right)\left(\frac{1}{I_0} - \frac{1}{I_1}\right) \quad \text{kip} \cdot \text{ft}^2$$

$$R'_B = \frac{wL^3}{24I_0} - \frac{wb^2}{24}\left(4 - \frac{3b}{L}\right)\left(\frac{1}{I_0} - \frac{1}{I_1}\right) \quad \text{kip} \cdot \text{ft}^2$$

$$M'_D = R'_B \frac{L}{2} - \text{area I} \frac{3L}{16}$$

$$= \frac{5wL^4}{384I_0} - \frac{wb^3}{48}(4L - 3b)\left(\frac{1}{I_0} - \frac{1}{I_1}\right) \quad \text{kip} \cdot \text{ft}^3$$

$$\Delta_D = \frac{5wL^4}{384EI_0} - \frac{wL^4}{48EI_0}(4p^3 - 3p^4)\left(\frac{n-1}{n}\right)$$

$$= \frac{wL^4}{EI_0}\left[0.013021 - f_2(p)\left(\frac{n-1}{n}\right)\right] \quad \text{in downward} \quad (5\text{-}8)$$

where

$$f_2(p) = \frac{1}{48}(4p^3 - 3p^4)$$

The reduction in deflection at the center of the beam due to the cover plates at the left end of the beam is $f_2(p)[(n-1)/n](wL^4/EI_0)$.

When the cover plates are at the right end of the beam, use p' and b' to represent the right-end cover plates.

$$\Delta_D = \frac{wL^4}{EI_0}\left[0.01302 - f_2(p')\left(\frac{n-1}{n}\right)\right] \quad \text{in downward} \quad (5\text{-}9)$$

The coefficients of deflection at the center of the beam for the cover plates at the right end are the same here as when the cover plates are at the left end of the beam.

A general equation for determining the deflection at the center of a simple-

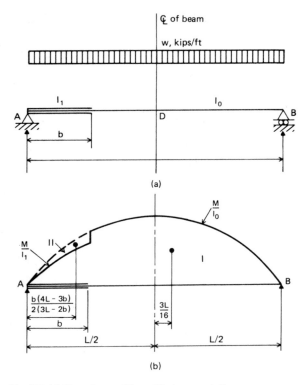

Fig. 5-8. (a) Given beam; (b) modified moment diagram.

support, cover-plated beam, carrying a uniformly distributed load over the whole span, is derived by combining Eqs. (5-7), (5-8), and (5-9)

$$\Delta_D = \frac{wL^4}{EI_0}\left\{0.013021 - [f_2(m) + f_2(p) + f_2(p')]\left(\frac{n-1}{n}\right)\right\} \quad \text{in downward} \quad (5\text{-}10)$$

where $f_2(m)$ = coefficient of deflection due to cover plates at the center of a simple-support beam carrying a uniformly distributed load over the whole span (Table 5-3)

$f_2(p)$ = coefficient of deflection due to cover plates at the left end of a simple-support beam carrying a uniformly distributed load over the whole span (Table 5-4)

$f_2(p')$ = coefficient of deflection due to cover plates at the right end of a simple-support beam carrying a uniformly distributed load over the whole span (Table 5-4)

Tables 5-3 and 5-4, at the end of the chapter, provide the coefficients $f_2(m)$, $f_2(p)$, and $f_2(p')$. In Eq. (5-10) it is assumed that all cover plates have the same ratio for the moments of inertia. Otherwise, the value of $(n - 1)/n$ would be different for each part. The cover plates are also assumed to consist of one set of plates. If there is more than one set of cover plates, additional ratios reflecting the moments of inertia of the additional plates should be used to obtain the additional reductions in deflection, as illustrated in Sec. 5-2.

Example 5-5

Find the deflection Δ_D for a uniformly distributed load over an entire span, as shown in Fig. 5-9, with $a = 0.6L$ ft at the center for the cover plates.

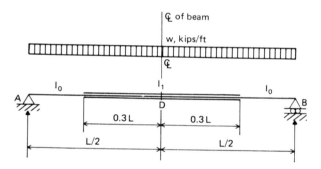

Fig. 5-9.

Solution

$$n = \frac{I_1}{I_0} = 1.5 \qquad \left(\frac{n-1}{n}\right) = 0.333$$

$$m = 0.6 \qquad f_2(m) = 0.011888 \text{ (from Table 5-3)}$$

From Eq. (5-10)

$$\Delta_D = \frac{wL^4}{EI_0}(0.013021 - 0.011888 \times 0.333)$$

$$= \frac{wL^4}{EI_0} 0.009062 \qquad \text{in downward} \qquad (ans.)$$

Example 5-6

The conditions are the same as in Example 5-5, but an additional set of cover plates with $a = 0.3L$ ft is added to the center, as shown in Fig. 5-10. Determine the deflection Δ_D.

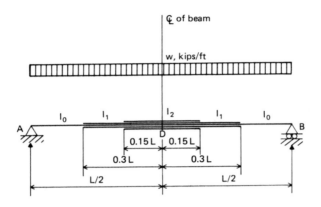

Fig. 5-10.

Solution

$$n_1 = \frac{I_1}{I_0} = 1.5 \qquad n_2 = \frac{I_2}{I_1} = 1.333 \qquad \left(\frac{n_2 - 1}{n_1 n_2}\right) = 0.167$$

$$m_2 = 0.3 \qquad f_2(m) = 0.007751 \text{ (from Table 5-3)}$$

From Eq. (5-10) and Example 5-5

$$\Delta_D = \frac{wL^4}{EI_0}(0.009062 - 0.007751 \times 0.167)$$

$$= \frac{wL^4}{EI_0} 0.007768 \quad \text{in downward} \qquad (ans.)$$

5-4 COVER-PLATED BEAM WITH END MOMENTS

For a cover-plated beam with end moments at both ends, the reduction in deflection is greater than for the same beam without the cover plates.

Cover Plates at the Center of the Beam

In order to derive a general equation for the deflection at the center of a beam due to the end moments M_{AB} and M_{BA}, with the cover plates at the center of the beam, we assume that $a = mL$, $m = 0$ to 1, and $n = I_1/I_0$. From Fig. 5-11(b) we can see that

$$\text{Area I} = \frac{L}{8I_0}(3M_{AB} + M_{BA}) \quad \text{kip} \cdot \text{ft}^2$$

$$\text{Area II} = \frac{L}{8I_0}(M_{AB} + 3M_{BA}) \quad \text{kip} \cdot \text{ft}^2$$

$$\text{Area III} = \frac{mL(M_{AB} + M_{BA})}{2}\left(\frac{1}{I_0} - \frac{1}{I_1}\right) \quad \text{kip} \cdot \text{ft}^2$$

$$\text{Area III}' = \frac{mL}{4}\left[\left(1 + \frac{m}{2}\right)M_{AB} + \left(1 - \frac{m}{2}\right)M_{BA}\right]\left(\frac{1}{I_0} - \frac{1}{I_1}\right) \quad \text{kip} \cdot \text{ft}^2$$

$$R'_A = \frac{L}{6I_0}(2M_{AB} + M_{BA}) - \frac{mL}{12}[(3 + m^2)M_{AB} + (3 - m^2)M_{BA}]$$

$$\times \left(\frac{1}{I_0} - \frac{1}{I_1}\right) \quad \text{kip} \cdot \text{ft}^2$$

$$R'_B = \frac{L}{6I_0}(2M_{BA} + M_{AB}) - \frac{mL}{12}[(3 - m^2)M_{AB} + (3 + m^2)M_{BA}]$$

$$\times \left(\frac{1}{I_0} - \frac{1}{I_1}\right) \quad \text{kip} \cdot \text{ft}^2$$

$$M'_D = R'_A \frac{L}{2} - \text{area I} \frac{L(5M_{AB} + M_{BA})}{6(3M_{AB} + M_{BA})} + \text{area III}'$$

$$\times \frac{(mL/2)[2h_1 + (M_{AB} + M_{BA})/2]}{3[h_1 + (M_{AB} + M_{BA})/2]}$$

$$= \frac{L^2}{16I_0}(M_{AB} + M_{BA}) - \frac{L^2}{16}(2m - m^2)(M_{AB} + M_{BA})$$

$$\times \left(\frac{1}{I_0} - \frac{1}{I_1}\right) \quad \text{kip} \cdot \text{ft}^3$$

$$\Delta'_D = -\frac{L^2}{16EI_0}(M_{AB} + M_{BA})\left[1 - (2 - m)m\left(\frac{n-1}{n}\right)\right]$$

$$= -\frac{L^2}{EI_0}(M_{AB} + M_{BA})\left[0.0625 - f_3(m)\left(\frac{n-1}{n}\right)\right] \quad \text{in upward}$$

(5-11)

where

$$f_3(m) = \frac{1}{16}(2 - m)m$$

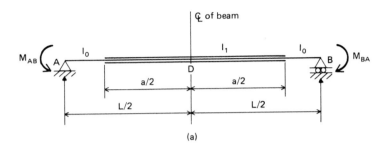

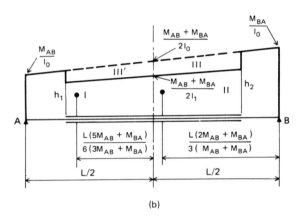

Fig. 5-11. (a) Given beam; (b) modified moment diagram.

Equation (5-11) gives the reduction in deflection at the center of the beam due to end moments, with cover plates at the center of the beam.

Cover Plates at the End of the Beam

In order to derive a general equation for the deflection at the center of a beam due to the end moments M_{AB} and M_{BA}, with the cover plates at the left end of the beam, we assume that $b = pL$, $p = 0$ to 0.5, and $n =$

I_1/I_0. From Figure 5-12(b) it can be seen that

$$\text{Area I} = \frac{L}{8I_0}(3M_{AB} + M_{BA}) \quad \text{kip} \cdot \text{ft}^2$$

$$\text{Area II} = \frac{L}{8I_0}(M_{AB} + 3M_{BA}) \quad \text{kip} \cdot \text{ft}^2$$

$$\text{Area III} = \frac{pL}{2}[(2-p)M_{AB} + pM_{BA}]\left(\frac{1}{I_0} - \frac{1}{I_1}\right) \quad \text{kip} \cdot \text{ft}^2$$

$$R'_A = \frac{L}{6I_0}(2M_{AB} + M_{BA}) - \frac{L}{6}[(2p^2 - 6p + 6)pM_{AB}$$
$$+ (3-2p)p^2 M_{BA}]\left(\frac{1}{I_0} - \frac{1}{I_1}\right) \quad \text{kip} \cdot \text{ft}^2$$

$$R'_B = \frac{L}{6I_0}(M_{AB} + 2M_{BA}) - \frac{L}{6}[(3-2p)p^2 M_{AB}$$
$$+ 2p^3 M_{BA}]\left(\frac{1}{I_0} - \frac{1}{I_1}\right) \quad \text{kip} \cdot \text{ft}^2$$

$$M'_D = R'_B \frac{L}{2} - \text{area II} \frac{L(M_{AB} + 5M_{BA})}{6(M_{AB} + 3M_{BA})}$$
$$= \frac{L^2}{16I_0}(M_{AB} + M_{BA}) - \frac{L^2}{12}[(3-2p)p^2 M_{AB} + 2p^3 M_{BA}]$$
$$\times \left(\frac{1}{I_0} - \frac{1}{I_1}\right) \quad \text{kip} \cdot \text{ft}^3$$

$$\Delta'_D = -\frac{L^2}{EI_0}\left\{\frac{1}{16}(M_{AB} + M_{BA}) - \left[\frac{p^2}{12}(3-2p)M_{AB} + \frac{p^3}{6}M_{BA}\right]\right.$$
$$\left. \times \left(\frac{n-1}{n}\right)\right\}$$

$$= -\frac{L^2}{EI_0}\left\{0.0625(M_{AB} + M_{BA}) - [f_{AB}(p)M_{AB} + f_{BA}(p)M_{BA}]\right.$$
$$\left. \times \left(\frac{n-1}{n}\right)\right\} \quad \text{in upward} \quad (5\text{-}12)$$

where

$$f_{AB}(p) = \frac{p^2}{12}(3-2p) \quad \text{for } M_{AB}$$

$$f_{BA}(p) = \frac{p^3}{6} \quad \text{for } M_{BA}$$

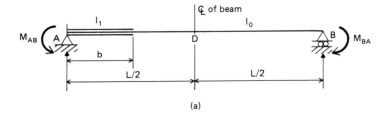

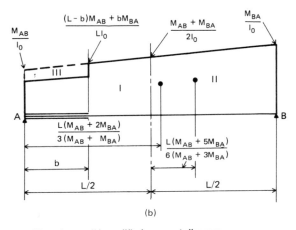

Fig. 5-12. (a) Given beam; (b) modified moment diagram.

Equation (5-12) gives the reduction in the deflection at the center of the beam due to the end moments, with the cover plates at the left end of the beam.

When the cover plates are at the right end of the beam, $f_{AB}(p)$ becomes $f_{BA}(p')$ and $f_{BA}(p)$ becomes $f_{AB}(p')$.

$$\Delta'_D = -\frac{L^2}{EI_0}\left\{0.0625(M_{AB} + M_{BA}) - [f_{AB}(p')M_{AB} + f_{BA}(p')M_{BA}]\left(\frac{n-1}{n}\right)\right\} \quad \text{in upward} \quad (5\text{-}13)$$

A general equation can be found by combining Eqs. (5-11), (5-12), and (5-13)

$$\Delta'_D = -\frac{L^2}{EI_0}\left\{0.0625(M_{AB} + M_{BA}) - [f_3(m)(M_{AB} + M_{BA}) + f_{AB}(p)M_{AB} + f_{BA}(p)M_{BA} + f_{AB}(p')M_{AB} + f_{BA}(p')M_{BA}]\left(\frac{n-1}{n}\right)\right\} \quad \text{in upward} \quad (5\text{-}14)$$

where $f_3(m)$ = coefficient of deflection of a beam due to the end moments, with cover plates at the center of the beam (Table 5-5)

$f_{AB}(p)$, $f_{BA}(p)$ = coefficients of deflection of a beam due to the end moments, with cover plates at the left end of the beam (Table 5-6)

$f_{AB}(p')$, $f_{BA}(p')$ = coefficients of deflection of a beam due to the end moments, with cover plates at the right end of the beam (Table 5-6)

Tables 5-5 and 5-6, at the end of the chapter, provide the coefficients of $f_3(m)$, $f_3(p)$, and $f_3(p')$. The cover plates are assumed to be the same as for Tables 5-1 through 5-4. In Eqs. (5-3), (5-7), and (5-11), the cover plates are placed at the center of the beam. When $m = 1$, the cross section of the entire beam changes from I_0 to I_1. Similarly, in Eqs. (5-5), (5-9), and (5-13), the cover plates are placed at the end of the beam. When $p = p' = 0.5$, the same result will be obtained, whether the plates are at the right or left end. The cover plates at the center of the beam are assumed to be symmetrical about the center. When the cover plates are not symmetrically placed at the center of the beam, use the end over plates to adjust for the asymmetrical part.

M_{AB} and M_{BA} are fixed-end moments. These moments vary with the varying cross sections of beams. The solution of these moments will be illustrated in Chapter 6.

TABLE 5-1 COEFFICIENTS OF DEFLECTION DUE TO COVER PLATES AT THE CENTER OF A SIMPLE-SUPPORT BEAM CARRYING A CONCENTRATED LOAD

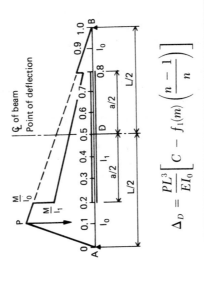

$$\Delta_D = \frac{PL^3}{EI_0}\left[C - f_1(m)\left(\frac{n-1}{n}\right)\right]$$

Extent of cover plates, $m = a/L$	Coefficient $f_1(m)$ for P at one-tenth points of beam								
	0.1L	0.2L	0.3L	0.4L	0.5L	0.6L	0.7L	0.8L	0.9L
0.025	0.000309	0.000617	0.000926	0.001234	0.001524	0.001234	0.000926	0.000617	0.000309
0.050	0.000609	0.001219	0.001828	0.002437	0.002971	0.002437	0.001828	0.001219	0.000609
0.075	0.000902	0.001805	0.002707	0.003609	0.004345	0.003609	0.002707	0.001805	0.000902
0.100	0.001187	0.002375	0.003562	0.004750	0.005646	0.004750	0.003562	0.002375	0.001187
0.125	0.001465	0.002930	0.004394	0.005859	0.006877	0.005859	0.004394	0.002930	0.001465
0.150	0.001734	0.003469	0.005203	0.006937	0.008039	0.006937	0.005203	0.003469	0.001734
0.175	0.001996	0.003992	0.005988	0.007984	0.009135	0.007984	0.005988	0.003992	0.001996
0.200	0.002250	0.004500	0.006750	0.009000	0.010167	0.009000	0.006750	0.004500	0.002250

0.225	0.002496	0.004992	0.007488	0.009969	0.011136	0.009969	0.007488	0.004992	0.002496
0.250	0.002734	0.005469	0.008203	0.010877	0.012044	0.010877	0.008203	0.005469	0.002734
0.275	0.002965	0.005930	0.008894	0.011727	0.012894	0.011727	0.008894	0.005930	0.002965
0.300	0.003187	0.006375	0.009562	0.012520	0.013687	0.012520	0.009562	0.006375	0.003187
0.325	0.003402	0.006805	0.010207	0.013259	0.014426	0.013259	0.010207	0.006805	0.003402
0.350	0.003609	0.007219	0.010828	0.013945	0.015112	0.013945	0.010828	0.007219	0.003609
0.375	0.003809	0.007617	0.011426	0.014580	0.015747	0.014580	0.011426	0.007617	0.003809
0.400	0.004000	0.008000	0.012000	0.015166	0.016333	0.015166	0.012000	0.008000	0.004000
0.425	0.004184	0.008367	0.012539	0.015706	0.016872	0.015706	0.012539	0.008367	0.004184
0.450	0.004359	0.008719	0.013033	0.016200	0.017367	0.016200	0.013033	0.008719	0.004359
0.475	0.004527	0.009055	0.013485	0.016652	0.017818	0.016652	0.013485	0.009055	0.004527
0.500	0.004687	0.009375	0.013895	0.017062	0.018229	0.017062	0.013895	0.009375	0.004687
0.525	0.004840	0.009680	0.014267	0.017433	0.018600	0.017433	0.014267	0.009680	0.004840
0.550	0.004984	0.009969	0.014601	0.017768	0.018935	0.017768	0.014601	0.009969	0.004984
0.575	0.005121	0.010242	0.014900	0.018067	0.019234	0.018067	0.014900	0.010242	0.005121
0.600	0.005250	0.010500	0.015166	0.018333	0.019500	0.018333	0.015166	0.010500	0.005250
0.625	0.005371	0.010734	0.015401	0.018567	0.019734	0.018567	0.015401	0.010734	0.005371
0.650	0.005484	0.010939	0.015606	0.018773	0.019940	0.018773	0.015606	0.010939	0.005484
0.675	0.005590	0.011117	0.015784	0.018951	0.020118	0.018951	0.015784	0.011117	0.005590
0.700	0.005687	0.011270	0.015937	0.019104	0.020271	0.019104	0.015937	0.011270	0.005687
0.725	0.005777	0.011399	0.016066	0.019233	0.020400	0.019233	0.016066	0.011399	0.005777
0.750	0.005859	0.011507	0.016174	0.019341	0.020507	0.019341	0.016174	0.011507	0.005859
0.775	0.005933	0.011595	0.016262	0.019429	0.020596	0.019429	0.016262	0.011595	0.005933
0.800	0.006000	0.011666	0.016333	0.019499	0.020666	0.019499	0.016333	0.011666	0.006000
0.825	0.006055	0.011721	0.016388	0.019554	0.020721	0.019554	0.016388	0.011721	0.006055
0.850	0.006096	0.011762	0.016429	0.019596	0.020763	0.019596	0.016429	0.011762	0.006096
0.875	0.006126	0.011792	0.016459	0.019625	0.020792	0.019625	0.016459	0.011792	0.006126
0.900	0.006146	0.011812	0.016478	0.019645	0.020812	0.019645	0.016478	0.011812	0.006146
0.925	0.006158	0.011824	0.016490	0.019657	0.020824	0.019657	0.016490	0.011824	0.006158
0.950	0.006164	0.011830	0.016497	0.019664	0.020830	0.019664	0.016497	0.011830	0.006164
0.975	0.006166	0.011832	0.016499	0.019666	0.020833	0.019666	0.016499	0.011832	0.006166
1.000	0.006167	0.011832	0.016499	0.019666	0.020833	0.019666	0.016499	0.011832	0.006167

TABLE 5-2 COEFFICIENTS OF DEFLECTION DUE TO COVER PLATES AT THE END OF A SIMPLE-SUPPORT BEAM CARRYING A CONCENTRATED LOAD

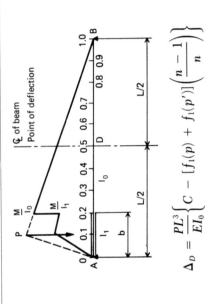

$$\Delta_D = \frac{PL^3}{EI_0}\left\{C - [f_1(p) + f_1(p')]\left(\frac{n-1}{n}\right)\right\}$$

Extent of cover plates at left end, $p = b/L$*	Coefficient $f_1(p)$ for P at one-tenth points of beam									
	0.1L	0.2L	0.3L	0.4L	0.5L	0.6L	0.7L	0.8L	0.9L
0.0125	0.000000	0.000000	0.000000	0.000000	0.000000	0.000000	0.000000	0.000000	0.000000
0.0250	0.000002	0.000002	0.000002	0.000002	0.000001	0.000001	0.000000	0.000001	0.000000
0.0375	0.000008	0.000007	0.000006	0.000005	0.000004	0.000004	0.000003	0.000002	0.000000
0.0500	0.000019	0.000017	0.000015	0.000013	0.000010	0.000008	0.000006	0.000004	0.000002
0.0625	0.000037	0.000033	0.000028	0.000024	0.000020	0.000016	0.000012	0.000008	0.000004
0.0750	0.000063	0.000056	0.000049	0.000042	0.000035	0.000028	0.000021	0.000014	0.000007
0.0875	0.000100	0.000089	0.000078	0.000067	0.000056	0.000045	0.000033	0.000022	0.000011
0.1000	0.000150	0.000133	0.000117	0.000100	0.000083	0.000067	0.000050	0.000033	0.000017
0.1125	0.000209	0.000190	0.000166	0.000142	0.000119	0.000095	0.000071	0.000047	0.000024
0.1250	0.000275	0.000260	0.000228	0.000195	0.000163	0.000130	0.000098	0.000065	0.000033
0.1375	0.000346	0.000347	0.000303	0.000260	0.000217	0.000173	0.000130	0.000087	0.000043
0.1500	0.000423	0.000450	0.000394	0.000337	0.000281	0.000225	0.000169	0.000112	0.000056

0.1625	0.000505	0.000572	0.000501	0.000429	0.000358	0.000286	0.000215	0.000143	0.000072
0.1750	0.000593	0.000715	0.000625	0.000536	0.000447	0.000357	0.000268	0.000179	0.000089
0.1875	0.000686	0.000879	0.000769	0.000659	0.000549	0.000439	0.000330	0.000220	0.000110
0.2000	0.000783	0.001067	0.000933	0.000800	0.000667	0.000533	0.000400	0.000267	0.000133
0.2125	0.000886	0.001271	0.001119	0.000960	0.000800	0.000640	0.000480	0.000320	0.000160
0.2250	0.000992	0.001485	0.001329	0.001139	0.000949	0.000759	0.000570	0.000380	0.000190
0.2375	0.001104	0.001707	0.001563	0.001340	0.001116	0.000893	0.000670	0.000447	0.000223
0.2500	0.001219	0.001937	0.001823	0.001563	0.001302	0.001042	0.000781	0.000521	0.000260
0.2625	0.001338	0.002176	0.002110	0.001809	0.001507	0.001206	0.000904	0.000603	0.000301
0.2750	0.001461	0.002421	0.002426	0.002080	0.001733	0.001386	0.001040	0.000693	0.000347
0.2875	0.001587	0.002674	0.002772	0.002376	0.001980	0.001584	0.001188	0.000792	0.000396
0.3000	0.001717	0.002933	0.003150	0.002700	0.002250	0.001800	0.001350	0.000900	0.000450
0.3125	0.001849	0.003199	0.003548	0.003052	0.002543	0.002035	0.001526	0.001017	0.000509
0.3250	0.001985	0.003470	0.003955	0.003433	0.002861	0.002289	0.001716	0.001144	0.000572
0.3375	0.002124	0.003747	0.004371	0.003844	0.003204	0.002563	0.001922	0.001281	0.000641
0.3500	0.002264	0.004029	0.004794	0.004287	0.003573	0.002858	0.002144	0.001429	0.000715
0.3625	0.002408	0.004316	0.005224	0.004763	0.003970	0.003176	0.002382	0.001588	0.000794
0.3750	0.002553	0.004607	0.005660	0.005273	0.004395	0.003516	0.002637	0.001758	0.000879
0.3875	0.002701	0.004901	0.006102	0.005819	0.004849	0.003879	0.002909	0.001939	0.000970
0.4000	0.002850	0.005200	0.006550	0.006400	0.005333	0.004267	0.003200	0.002133	0.001067
0.4125	0.003001	0.005501	0.007002	0.010523	0.005849	0.004679	0.003509	0.002340	0.001170
0.4250	0.003153	0.005805	0.007458	0.011131	0.006397	0.005118	0.003838	0.002559	0.001279
0.4375	0.003306	0.006112	0.007918	0.011744	0.006978	0.005583	0.004187	0.002797	0.001396
0.4500	0.003460	0.006421	0.008381	0.012361	0.007594	0.006075	0.004556	0.003037	0.001519
0.4625	0.003615	0.006731	0.008846	0.012981	0.008244	0.006595	0.004947	0.003298	0.001649
0.4750	0.003771	0.007042	0.009313	0.013604	0.008931	0.007145	0.005359	0.003572	0.001786
0.4875	0.003927	0.007354	0.009781	0.014228	0.009655	0.007724	0.005793	0.003862	0.001931
0.5000	0.004083	0.007666	0.010250	0.014853	0.010417	0.008333	0.006250	0.004167	0.002083

*When the cover plates are at the right end, read $f_1(p')$ at the complement of the load position being considered.

TABLE 5-3 COEFFICIENTS OF DEFLECTION DUE TO COVER PLATES AT THE CENTER OF A SIMPLE-SUPPORT BEAM CARRYING A UNIFORMLY DISTRIBUTED LOAD OVER THE WHOLE SPAN

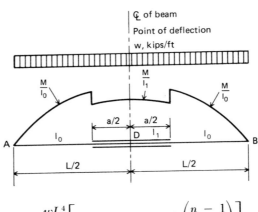

$$\Delta_D = \frac{wL^4}{EI_0}\left[0.013021 - f_2(m)\left(\frac{n-1}{n}\right)\right]$$

Extent of cover plates, $m = a/L$	Coefficient $f_2(m)$	Extent of cover plates, $m = a/L$	Coefficient $f_2(m)$
0.025	0.000771	0.525	0.011186
0.050	0.001522	0.550	0.011443
0.075	0.002251	0.575	0.011676
0.100	0.002959	0.600	0.011888
0.125	0.003644	0.625	0.012077
0.150	0.004305	0.650	0.012245
0.175	0.004942	0.675	0.012393
0.200	0.005554	0.700	0.012522
0.225	0.006142	0.725	0.012632
0.250	0.006704	0.750	0.012726
0.275	0.007240	0.775	0.012804
0.300	0.007751	0.800	0.012867
0.325	0.008235	0.825	0.012917
0.350	0.008694	0.850	0.012954
0.375	0.009127	0.875	0.012982
0.400	0.009533	0.900	0.013001
0.425	0.009914	0.925	0.013012
0.450	0.010269	0.950	0.013018
0.475	0.010599	0.975	0.013020
0.500	0.010905	1.000	0.013021

TABLE 5-4 COEFFICIENTS OF DEFLECTION DUE TO COVER PLATES AT THE END OF A SIMPLE-SUPPORT BEAM CARRYING A UNIFORMLY DISTRIBUTED LOAD OVER THE WHOLE SPAN

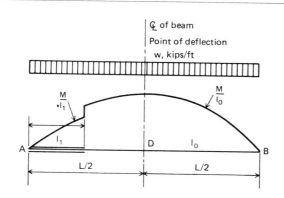

$$\Delta_D = \frac{wL^4}{EI}\left\{0.013021 - [f_2(p) + f_2(p')]\left(\frac{n-1}{n}\right)\right\}$$

Extent of cover plates at left end, $p = b/L$*	Coefficient $f_2(p)$	Extent of cover plates at left end, $p = b/L$*	Coefficient $f_2(p)$
0.0125	0.000000	0.2625	0.001211
0.0250	0.000001	0.2750	0.001376
0.0375	0.000004	0.2875	0.001553
0.0500	0.000010	0.3000	0.001744
0.0625	0.000019	0.3125	0.001947
0.0750	0.000033	0.3250	0.002163
0.0875	0.000052	0.3375	0.002393
0.0100	0.000077	0.3500	0.002635
0.1125	0.000109	0.3625	0.002890
0.1250	0.000148	0.3750	0.003159
0.1375	0.000194	0.3875	0.003439
0.1500	0.000249	0.4000	0.003733
0.1625	0.000314	0.4125	0.004039
0.1750	0.000388	0.4250	0.004358
0.1875	0.000472	0.4375	0.004689
0.2000	0.000567	0.4500	0.005031
0.2125	0.000672	0.4625	0.005385
0.2250	0.000789	0.4750	0.005749
0.2375	0.000918	0.4875	0.006125
0.2500	0.001058	0.5000	0.006510

*When the cover plates are at the right end, read the same values for $f_2(p')$ as for $f_2(p)$.

TABLE 5-5 COEFFICIENTS OF DEFLECTION OF A BEAM DUE TO END MOMENTS—COVER PLATES AT THE CENTER OF THE BEAM

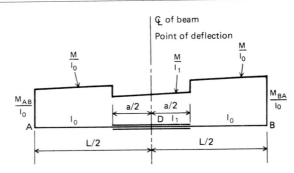

$$\Delta_D'' = -\frac{L^2}{EI}\left[0.0625(M_{AB} + M_{BA}) - f_3(m)(M_{AB} + M_{BA})\left(\frac{n-1}{n}\right)\right]$$

Extent of cover plates, $m = a/L$	Coefficient $f_3(m)$ for $M_{AB} + M_{BA}$	Extent of cover plates, $m = a/L$	Coefficient $f_3(m)$ for $M_{AB} + M_{BA}$
0.025	0.003086	0.525	0.048398
0.050	0.006094	0.550	0.049844
0.075	0.009023	0.575	0.051211
0.100	0.011875	0.600	0.052500
0.125	0.014648	0.625	0.053711
0.150	0.017344	0.650	0.054844
0.175	0.019961	0.675	0.055898
0.200	0.022500	0.700	0.056875
0.225	0.024961	0.725	0.057773
0.250	0.027344	0.750	0.058594
0.275	0.029648	0.775	0.059336
0.300	0.031875	0.800	0.060000
0.325	0.034023	0.825	0.060586
0.350	0.036094	0.850	0.061094
0.375	0.038086	0.875	0.061523
0.400	0.040000	0.900	0.061875
0.425	0.041836	0.925	0.062148
0.450	0.043594	0.950	0.062344
0.475	0.045273	0.975	0.062461
0.500	0.046875	1.000	0.062500

TABLE 5-6 COEFFICIENTS OF DEFLECTION OF A BEAM DUE TO END MOMENTS—COVER PLATES AT THE END OF THE BEAM

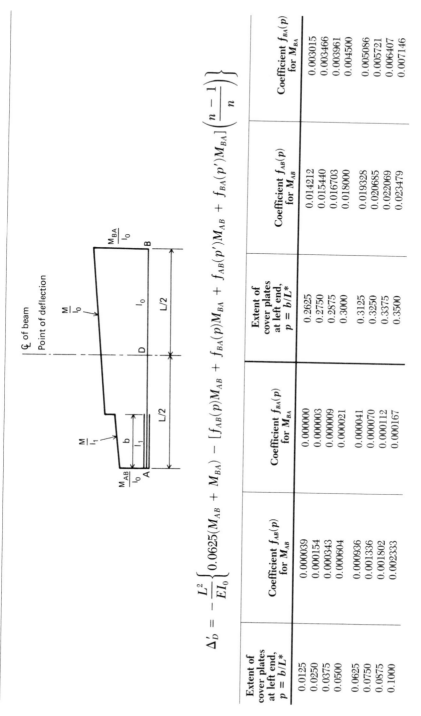

$$\Delta'_D = -\frac{L^2}{EI_0}\left\{0.0625(M_{AB} + M_{BA}) - [f_{AB}(p)M_{AB} + f_{BA}(p)M_{BA} + f_{AB}(p')M_{AB} + f_{BA}(p')M_{BA}]\left(\frac{n-1}{n}\right)\right\}$$

Extent of cover plates at left end, $p = b/L^*$	Coefficient $f_{AB}(p)$ for M_{AB}	Coefficient $f_{BA}(p)$ for M_{BA}	Extent of cover plates at left end, $p = b/L^*$	Coefficient $f_{AB}(p)$ for M_{AB}	Coefficient $f_{BA}(p)$ for M_{BA}
0.0125	0.000039	0.000000	0.2625	0.014212	0.003015
0.0250	0.000154	0.000003	0.2750	0.015440	0.003466
0.0375	0.000343	0.000009	0.2875	0.016703	0.003961
0.0500	0.000604	0.000021	0.3000	0.018000	0.004500
0.0625	0.000936	0.000041	0.3125	0.019328	0.005086
0.0750	0.001336	0.000070	0.3250	0.020685	0.005721
0.0875	0.001802	0.000112	0.3375	0.022069	0.006407
0.1000	0.002333	0.000167	0.3500	0.023479	0.007146

TABLE 5-6 CONTINUED

Extent of cover plates at left end, $p = b/L$*	Coefficient $f_{AB}(p)$ for M_{AB}	Coefficient $f_{BA}(p)$ for M_{BA}	Extent of cover plates at left end, $p = b/L$*	Coefficient $f_{AB}(p)$ for M_{AB}	Coefficient $f_{BA}(p)$ for M_{BA}
0.1125	0.002927	0.000237	0.3625	0.024912	0.001939
0.1250	0.003581	0.000326	0.3750	0.026367	0.008789
0.1375	0.004293	0.000433	0.3875	0.027841	0.009698
0.1500	0.005063	0.000563	0.4000	0.029333	0.010667
0.1625	0.005886	0.000715	0.4125	0.030841	0.011698
0.1750	0.006763	0.000893	0.4250	0.032362	0.012794
0.1875	0.007690	0.001099	0.4375	0.033895	0.013957
0.2000	0.008667	0.001333	0.4500	0.035438	0.015188
0.2125	0.009690	0.001599	0.4625	0.036988	0.016489
0.2250	0.010758	0.001898	0.4750	0.038544	0.017862
0.2375	0.011869	0.002233	0.4875	0.040104	0.019310
0.2500	0.013021	0.002604	0.5000	0.041667	0.020833

*When the cover plates are at the right end, read the table from the bottom up.

6
Deflection of Cover-Plated Continuous Beams

COVER PLATES ON CONTINUOUS BEAMS

The effect of cover plates on the deflection of a beam has been discussed in Chapter 5. The procedure for calculating the deflection of a continuous beam with cover plates is the same as that for calculating the deflection of a continuous beam without cover plates. By calculating the deflection due to load on a simple-support beam with the cover plates on the beam and the deflection of the same cover-plated beam due to the end moments, we can obtain the deflection of a cover-plated continuous beam. Equations (5-6), (5-10), and (5-14) are provided for this purpose.

However, the solution of the support moments for the cover-plated beam is different from the solution of the support moments for the non-cover-plated beam. Because the cover-plated beam has varying cross sections, the fixed-end moments, the stiffnesses, and the carryover factors must be determined for all the different cross sections before the support moments can be computed.

FIXED-END MOMENTS, STIFFNESSES, AND CARRYOVER FACTORS

Calculation of the fixed-end moments, stiffnesses, and carryover factors for members with variable cross sections is illustrated in this section.

Figure 6-1 shows a beam with cover plates at both ends. A concentrated load P is placed at any point on the beam. The fixed-end moment M_{AB} can be expressed as[1]

$$M_{AB} = P \frac{\int_0^L \frac{x\,dx}{I} \int_a^L \frac{x(x-a)\,dx}{I} - \int_0^L \frac{x^2\,dx}{I} \int_a^L \frac{(x-a)\,dx}{I}}{\int_0^L \frac{dx}{I} \int_0^L \frac{x^2\,dx}{I} - \left(\int_0^L \frac{x\,dx}{I}\right)^2} \qquad (6\text{-}1)$$

[1] Paul Andersen, *Statically Indeterminate Structures*, Ronald Press, New York, 1953, p. 191.

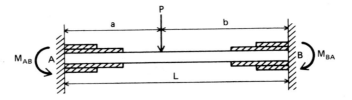

Fig. 6-1.

A similar expression for M_{BA} can be obtained by changing the symbol a to b. Since I is variable, integrations should be performed for each part of the cross section.

For a uniformly distributed load over the entire span of a beam with unsymmetrically variable cross sections, the fixed-end moment M_{AB} can be expressed as

$$M_{AB} = \frac{1}{2} wx \frac{\int_0^L \frac{x\,dx}{I} \int_0^L \frac{x^3\,dx}{I} - \left(\int_0^L \frac{x^2\,dx}{I}\right)^2}{\int_0^L \frac{dx}{I} \int_0^L \frac{x^2\,dx}{I} - \left(\int_0^L \frac{x\,dx}{I}\right)^2} \qquad (6\text{-}2)$$

When I is variable, the fixed-end moment M_{BA} can be found in the same manner by performing the integrations in the opposite direction.

If the beam has a symmetrically variable cross section, the two fixed-end moments for a uniformly distributed load become identical.

$$M_{AB} = M_{BA} = \frac{1}{2} wx \frac{L \int_0^L \frac{x\,dx}{I} - \int_0^L \frac{x^2\,dx}{I}}{\int_0^L \frac{dx}{I}} \qquad (6\text{-}3)$$

If the cross section is constant, Eq. (6-1) becomes $M_{AB} = Pab^2/L^2$ and $M_{BA} = Pa^2b/L^2$, while Eq. (6-2) becomes $M_{AB} = M_{BA} = \frac{1}{12}wL^2$.

Rotational stiffness can be found from

$$K = \frac{E \int_0^L \frac{x^2\,dx}{I}}{\int_0^L \frac{dx}{I} \int_0^L \frac{x^2\,dx}{I} - \left(\int_0^L \frac{x\,dx}{I}\right)^2} \qquad (6\text{-}4)$$

If the cross section is constant, Eq. (6-4) becomes $K = 4EI/L$. When the other end is simply supported or hinged, instead of being fixed, then $K = 3EI/L$.

Carryover factors can be found from

$$C = \frac{L \int_0^L \frac{x\,dx}{I} - \int_0^L \frac{x^2\,dx}{I}}{\int_0^L \frac{x^2\,dx}{I}} \tag{6-5}$$

Integrating and solving for C will yield the carryover factor from one end to the other. If the cross section is constant, when integrating from zero to L, C will become ½.

Lateral stiffness can be found from

$$M_{AB} = \frac{\Delta}{L} K_{AB}(1 + C_{AB}) \quad \text{and} \quad M_{BA} = \frac{\Delta}{L} K_{BA}(1 + C_{BA}) \tag{6-6}$$

If the cross section is constant, the fixed-end moment will be $M = 6EI\Delta/L^2$. Likewise, if one end is simply supported, the fixed-end moment at the other end will be $M = 3EI\Delta/L^2$. This is half the stiffness value of the member having both ends fixed.

In computing the fixed-end moments, stiffnesses, and carryover factors for members with variable cross sections, the column-analogy method also can be employed.[2] However, both the integration method and the column-analogy method are laborious. In order to simplify the calculation, R. A. Caughy, a professor of civil engineering at Iowa State College in Ames, Iowa, and Richard Sigmund Cebula, a former head of the engineering department at St. Martin's College in Olympia, Washington, have presented a series of 36 charts for determining the fixed-end moments, stiffnesses, and carryover factors for members with variable cross sections when cover plates are used. Using these charts, the designer can readily find, for each condition, the coefficients which make possible the calculation of the final end moments at the supports. Thus, these charts (presented in Appendix B) are very useful in computing the deflection of cover-plated continuous beams.

The charts are made for beams with the cover plates placed at the end. However, when the cover plates are placed at the center of the beam, the charts can still be applied. First, from the charts find the values of a beam cover-plated at both ends. By subtracting these values from those for a fully cover-plated beam, the values of the complementary center-cover-plated beam can be found. Examples 6-5 and 6-6 illustrate this procedure.

[2]Chu-kia Wang, *Statically Indeterminate Structures*, McGraw-Hill, New York, 1953, p. 298.

Example 6-1

Determine the deflection Δ_D for the beam shown in Fig. 6-2, with $p = 0.25$, $n = I_1/I_0 = 2$, and $(n - 1)/n = 1/2$.

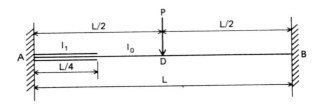

Fig. 6-2.

Solution

The values for the fixed-end moments are

$M_{AB} = 0.167PL$ (from Chart 20) $M_{BA} = 0.108PL$ (from Chart 29)

$f_{AB}(p) = 0.013021$ for M_{AB} (from Table 5-6)

$f_{BA}(p) = 0.002604$ for M_{BA} (from Table 5-6)

By Eq. (5-14)

$$\Delta_D' = -\frac{L^2}{EI_0}\left[0.0625(0.167PL + 0.108PL)\right.$$
$$\left. - (0.013021 \times 0.167PL + 0.002604 \times 0.108PL)\frac{1}{2}\right]$$

$$= -\frac{PL^3}{EI_0}0.015960 \text{ in upward}$$

The deflection of a simple-support, cover-plated beam (from Example 5-4) is

$$\Delta_D = \frac{PL^3}{EI_0}0.020182 \text{ in downward}$$

Therefore, the deflection of the beam in Fig. 6-2 is

$$\Delta_D = \frac{PL^3}{EI}(0.020182 - 0.015960)$$

$$= \frac{PL^3}{EI_0}0.004222 \text{ in downward} \qquad (ans.)$$

Example 6-2

Determine the deflection Δ_D for the beam shown in Fig. 6-3, with $p = p' = 0.25$, $n = I_1/I_0 = 2$, and $(n - 1)/n = 1/2$.

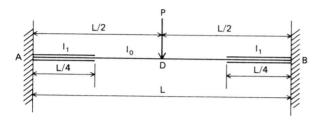

Fig. 6-3.

Solution

The values for the fixed-end moments are

$$M_{AB} = M_{BA} = 0.1458PL \quad \text{(from Chart 11)}$$
$$f_{AB} = 0.013021 \quad \text{for } M_{AB} \text{ (from Table 5-6)}$$
$$f_{BA} = 0.002604 \quad \text{for } M_{BA} \text{ (from Table 5-6)}$$

By Eq. (5-14)

$$\Delta'_D = -\frac{L^2}{EI_0}\left[0.0625 \times 2 \times 0.1458PL\right.$$
$$\left. - (0.013021 + 0.002604)2 \times 0.1458 \times \frac{1}{2}\right]$$
$$= -\frac{PL^3}{EI_0} 0.015947 \text{ in upward}$$

For the deflection of a simple-support beam without cover plates, $C = 0.020833$ (from Table 2-1). For the same beam with cover plates at both ends, $f_1(p) = f_1(p') = 0.001302$ (from Table 5-2). By Eq. (5-6), the deflection for the simple-support beam with cover plates at both ends is given by

$$\Delta_D = \frac{PL^3}{EI_0}\left(0.020833 - 2 \times 0.001302 \times \frac{1}{2}\right)$$
$$= \frac{PL^3}{EI_0} 0.019531 \text{ in downward}$$

Therefore, the deflection of the beam in Fig. 6-3 is

$$\Delta_D = \frac{PL^3}{EI_0}(0.019531 - 0.015947)$$

$$= \frac{PL^3}{EI_0} 0.003684 \text{ in downward} \qquad (ans.)$$

Example 6-3
Determine the deflection Δ_D for the beam shown in Fig. 6-4, with $p = 0.25$, $n = I_1/I_0 = 2$, and $(n-1)/n = 1/2$.

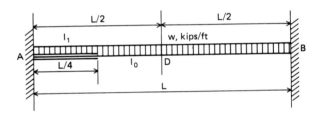

Fig. 6-4.

Solution
The values for the fixed-end moments are

$$M_{AB} = 0.1045wL^2 \qquad \text{(from Chart 34)}$$
$$M_{BA} = 0.0740wL^2 \qquad \text{(from Chart 36)}$$
$$f_{AB}(p) = 0.013021 \qquad \text{for } M_{AB} \text{ (from Table 5-6)}$$
$$f_{BA}(p) = 0.002604 \qquad \text{for } M_{BA} \text{ (from Table 5-6)}$$

By Eq. (5-14)

$$\Delta'_D = -\frac{L^2}{EI_0}\bigg[0.0625(0.1045 + 0.074)wL^2$$
$$- (0.013021 \times 0.1405wL^2 + 0.002604 \times 0.0740wL^2)\frac{1}{2}\bigg]$$

$$= -\frac{wL^4}{EI_0} 0.010380 \text{ in upward}$$

For the deflection of a simple-support beam without cover plates, $C = 0.013021$ (from Table 2-3). For the same beam with cover plates at the left end, $f_1(p) = 0.001058$ (from Table 5-4). By Eq. (5-10), the deflection for the simple-support beam with cover plates at the left end is given by

$$\Delta_D = \frac{wL^4}{EI_0}\left(0.013021 - 0.001058 \times \frac{1}{2}\right)$$

$$= \frac{wL^4}{EI_0} 0.012492 \text{ in downward}$$

Therefore, the deflection of the beam in Fig. 6-4 is

$$\Delta_D = \frac{wL^4}{EI_0}(0.012492 - 0.010380)$$

$$= \frac{wL^4}{EI_0} 0.002112 \text{ in downward} \quad\quad (ans.)$$

Example 6-4

Determine the deflection Δ_D for the beam shown in Fig. 6-5, with $p = p' = 0.25$, $n = I_1/I_0 = 2$, and $(n-1)/n = 1/2$.

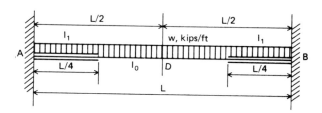

Fig. 6-5.

Solution

The values for the fixed-end moments are

$$M_{AB} = M_{BA} = 0.0938wL^2 \quad \text{(from Chart 35)}$$
$$f_{AB}(p) = 0.013021 \quad \text{for } M_{AB} \text{ (from Table 5-6)}$$
$$f_{BA}(p) = 0.002604 \quad \text{for } M_{BA} \text{ (from Table 5-6)}$$

154 Chapter 6

By Eq. (5-14)

$$\Delta'_D = -\frac{L^2}{EI_0}\left[0.0625 \times 2 \times 0.0938wL^2 - (0.013021 + 0.002604)\right.$$
$$\left. \times 2 \times 0.0938wL^2 \times \frac{1}{2}\right]$$

$$= -\frac{wL^4}{EI_0} 0.010259 \text{ in upward}$$

For the deflection of a simple-support beam without cover plates, $C = 0.013021$ (from Table 2-3). For the same beam with cover plates at both ends, $f_1(p) = f_1(p') = 0.001058$ (from Table 5-4). By Eq. (5-10), the deflection of the simple-support beam with cover plates at both ends is given by

$$\Delta_D = \frac{wL^4}{EI_0}\left(0.013021 - 0.001058 \times 2 \times \frac{1}{2}\right)$$

$$= \frac{wL^4}{EI_0} 0.011963 \text{ in downward}$$

Therefore, the deflection of the beam in Fig. 6-5 is

$$\Delta_D = \frac{wL^4}{EI_0}(0.011963 - 0.010259)$$

$$= \frac{wL^4}{EI_0} 0.001704 \text{ in downward} \qquad (ans.)$$

Example 6-5

Determine the fixed-end moments M_{AB} and M_{BA} and the deflection Δ_D for the beam shown in Fig. 6-6, with $m = 0.50$, $n = I_1/I_0 = 2$, and $(n - 1)/n = 1/2$.

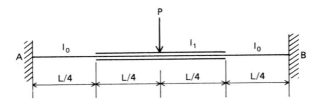

Fig. 6-6.

Solution

Find the fixed-end moments for the beam cover-plated at both ends.

$$p = p' = 0.25 \quad \text{and} \quad n = I_1/I_0 = 2$$

$$M_{AB} = M_{BA} = 0.1458PL \quad \text{kip} \cdot \text{ft (from Chart 11)}$$

For a fully cover-plated beam

$$M_{AB} = M_{BA} = 0.125PL \quad \text{kip} \cdot \text{ft (from Table 3-1)}$$

Therefore, the fixed-end moments of the beam in Fig. 6-6 are

$$M_{AB} = M_{BA} = 0.125PL - (0.1458 - 0.125)PL$$

$$= 0.1042PL \quad \text{kip} \cdot \text{ft} \quad (ans.)$$

By Eq. (5-14)

$$\Delta'_D = -\frac{L^2}{EI_0}\left(0.0625 \times 0.1042PL \times 2 - 0.046875\right.$$

$$\left. \times 0.01042PL \times 2 \times \frac{1}{2}\right)$$

$$= -\frac{PL^3}{EI_0} 0.008141 \text{ in upward}$$

The deflection for a simple-support, cover-plated beam (from Example 5-3) is

$$\Delta_D = \frac{PL^3}{EI_0} 0.011719 \text{ in downward}$$

Therefore, the deflection of the beam in Fig. 6-6 is

$$\Delta_D = \frac{PL^3}{EI_0}(0.011719 - 0.008141)$$

$$= \frac{PL^3}{EI_0} 0.003578 \text{ in downward} \quad (ans.)$$

Example 6-6

Determine the fixed-end moments M_{AB} and M_{BA} and the deflection Δ_D for the beam shown in Fig. 6-7, with $m = 0.50$, $n = I_1/I_0 = 2$, and $(n - 1)/n = 1/2$.

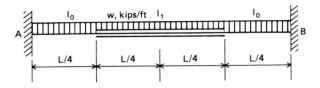

Fig. 6-7.

Solution

Find the fixed-end moments for the beam cover-plated at both ends.

$$p = p' = 0.25 \quad \text{and} \quad n = I_1/I_0 = 2$$

$$M_{AB} = M_{BA} = 0.09374wL^2 \quad \text{kip} \cdot \text{ft (from Chart 35)}$$

For a fully cover-plated beam

$$M_{AB} = M_{BA} = 0.083333wL^2 \quad \text{kip} \cdot \text{ft (from Table 3-1)}$$

Therefore, the fixed-end moments of the beam in Fig. 6-7 are

$$M_{AB} = M_{BA} = 0.083333wL^2 - (0.09374 - 0.083333)wL^2$$
$$= 0.07293wL^2 \quad \text{kip} \cdot \text{ft} \quad (ans.)$$

By Eq. (5-14)

$$\Delta'_D = -\frac{L^2}{EI_0}\bigg[0.0625 \times 0.07293wL^2 \times 2 - 0.046875$$
$$\times 0.07293wL^2 \times 2 \times \frac{1}{2}\bigg]$$

$$= -\frac{wL^4}{EI_0} 0.005598 \quad \text{in upward}$$

For the deflection of a simple-support beam without cover plates, $C = 0.013021$ (from Table 2-3). For the same beam with cover plates at the center, $f_2(m) = 0.010905$ (Table 5-3). By Eq. (5-10), the deflection for the simple-support beam with cover plates in the center is given by

$$\Delta_D = \frac{wL^4}{EI_0}\left(0.013021 - 0.010905 \times \frac{1}{2}\right)$$

$$= \frac{wL^4}{EI_0} 0.007568 \quad \text{in downward}$$

Therefore, the deflection for the beam in Fig. 6-7 is

$$\Delta_D = \frac{wL^4}{EI_0}(0.007568 - 0.005598)$$

$$= \frac{wL^4}{EI_0} 0.001970 \text{ in downward} \qquad (ans.)$$

From the above examples, it is seen that the fixed-end moment developed at one end depends not only on the condition at that end, but also on the condition at the other. Examples 6-1 and 6-3 demonstrate that when the cover plates are placed at one end, the fixed-end moment increases at that end and decreases at the uncovered end. Examples 6-2 and 6-4 show that when the cover plates are placed at both ends, the fixed-end moments at both ends increase. In contrast, when the cover plates are placed at the center of the beam, as in Examples 6-5 and 6-6, the fixed-end moments at both ends decrease. Generally speaking, for a beam having end moments at both ends, the reduction of deflection is greater if the beam has cover plates.

3 APPLICATION OF MOMENT DISTRIBUTION FOR COVER-PLATED CONTINUOUS BEAMS

As we have seen, the moment distribution begins with relative stiffnesses, carryover factors, and fixed-end moments and ends with the final end moments or support moments. Therefore, before starting the moment distribution, we should check that all values of these factors are correct and, when the moment-distribution table is finished, check that the final end moments obtained are balanced at each joint.

The procedure for finding the final moments at the supports for a continuous beam with variable cross section is as follows:

1. Determine the moments of inertia at each variable cross section and obtain their ratios, $n_1 = I_1/I_0$, $n_2 = I_2/I_1$, $n_3 = I_3/I_2$, ...

2. Determine the rotational stiffnesses, carryover factors, and fixed-end moments from the charts (Appendix B).

3. Check that the values of $C_{AB}K_{AB}$ and $C_{BA}K_{BA}$ are equal before starting the moment distribution.

4. When the moment-distribution table is finished, check that the final support moments obtained are balanced at each joint.

TABLE 6-1 VALUES OF k FOR ADJUSTED STIFFNESS FACTORS FOR THE MEMBERS WITH VARIABLE CROSS SECTIONS

Far end	Moment relation	Bending moment diagram	Unsymmetrically haunched, $C_{AB} \neq C_{BA}$	Symmetrically haunched, $C_{AB} = C_{BA}$
Fixed	$M_{BA} = C_{AB} M_{AB}$		$k = 1$	$k = 1$
Simple-support	$M_{BA} = 0$		$k = 1 - C_{AB} C_{BA}$	$k = 1 - C_{AB}^2$
With equal negative moment	$M_{BA} = M_{AB}$		$k = \dfrac{1 - C_{AB} C_{BA}}{1 + C_{BA}}$	$k = 1 - C_{AB}$
With equal moment	$M_{BA} = M_{AB}$		$k = \dfrac{1 - C_{AB} C_{BA}}{1 - C_{BA}}$	$k = 1 + C_{AB}$

Note: C = carryover factor; k = stiffness coefficient.

5. Compute the deflection at the center of the span step by step as illustrated in the examples.

Table 6-1 is useful in determining adjusted stiffness factors for the members with variable cross sections.

Example 6-7

A continuous beam has spans AB, BC, and CD and is loaded with two equal concentrated loads of P kips, as is shown in Fig. 6-8. Determine the deflection $\Delta_{0.5}$ in terms of PL_1^3/EI_0. Let $L_1:L_2:L_3 = 1:1.333:1$.

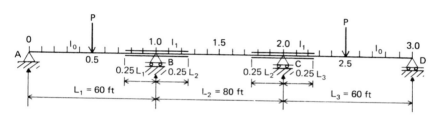

Fig. 6-8.

Solution

First, determine the stiffnesses, carryover factors, and fixed-end moments for the beam, with $p = p' = 0.25$, $n = I_1/I_0 = 1.39$, and $(n-1)/n = 0.281$

$C_{AB} = 0.571$ (Chart 5) $K_{AB} = 4.15I_0/L_1$ (Chart 2) $C_{AB}K_{AB} = 2.370$

$C_{BA} = 0.480$ (Chart 6) $K_{BA} = 4.94I_0/L_1$ (Chart 3) $C_{BA}K_{BA} = 2.370$

$C_{BC} = 0.547$ (Chart 4) $K_{BC} = 5.15I_0/L_2 = 3.86I_0/L_1$ (Chart 1),

$$M_{AB} = 0.117PL_1 \text{ (Chart 29)} \quad M_{BC} = 0$$
$$M_{BA} = 0.143PL_1 \text{ (Chart 20)} \quad M_{CB} = 0$$

The modified K values are

$$K'_{BA} = (1 - C_{AB}C_{BA})K_{BA} = (1 - 0.571 \times 0.480) \times 4.94\frac{I_0}{L_1} = 3.586\frac{I_0}{L_1}$$

$$K'_{BC} = (1 - C_{BC})K_{BC} = (1 - 0.547) \times 3.86\frac{I_0}{L_1} = 1.749\frac{I_0}{L_1}$$

The moment-distribution information for Example 6-7 is given in the accompanying table.

Joint	A	B		C		D
Member	AB	BA	BC	CB	CD	DC
K	4.15	4.94	3.86	3.86	4.94	4.15
Modified K	—	3.586	1.749	1.749	3.586	—
Moment-distribution factor	1.00	0.672	0.328	0.328	0.672	1.00
Carryover factor	0.571	0.480	0.547	0.547	0.480	0.571
Fixed-end moment* balance	+11.70 −11.70	−14.30 + 9.61	— +4.69	— −4.69	+14.30 − 9.61	−11.70 +11.70
Carryover balance	+ 4.61 − 4.61	− 6.68 + 6.22	−2.57 +3.03	+2.57 −3.03	+ 6.68 − 6.22	− 4.61 + 4.61
Carryover balance	+ 2.99 − 2.99	− 2.63 + 2.88	−1.66 +1.41	+1.66 −1.41	+ 2.63 − 2.88	− 2.99 + 2.99
Carryover balance	+ 1.38 − 1.38	− 1.71 + 1.67	−0.77 +0.81	+0.77 −0.81	+ 1.71 − 1.67	− 1.38 + 1.38
Carryover balance	+ 0.80 − 0.80	− 0.79 + 0.83	−0.44 +0.40	+0.44 −0.40	+ 0.79 − 0.83	− 0.80 + 0.80
Carryover balance	+ 0.40 − 0.40	− 0.46 + 0.46	−0.22 +0.22	+0.22 −0.22	+ 0.46 − 0.46	− 0.40 + 0.40
Carryover balance	+ 0.22 − 0.22	− 0.23 + 0.24	−0.12 +0.11	+0.12 −0.11	+ 0.23 − 0.24	− 0.22 + 0.22
Total†	0	− 4.89	+4.89	−4.89	+ 4.89	0

*Multiply figures by $PL_1/100$ to get the fixed-end moment.
†$M = (PL_1/100) \times$ Total.

Compute the deflection at the center of span AB with the cover plates at the right end.

$$M_{AB} = 0 \quad M_{BA} = -0.0489 PL_1 \quad \text{kip} \cdot \text{ft}$$
$$C = 0.020833 \text{ (from Table 2-1)}$$
$$f_1(p') = 0.001302 \text{ (from Table 5-2)}$$
$$f_{BA}(p') = 0.013021 \text{ (from Table 5-6)}$$

The deflection for a simple-support, cover-plated beam is, by Eq. (5-6),

$$\Delta_{0.5} = \frac{PL_1^3}{EI_0}(0.020833 - 0.001302 \times 0.281)$$

$$= \frac{PL_1^3}{EI_0} 0.020467 \text{ in downward}$$

The reduction in deflection effected by the end moments is, by Eq. (5-14),

$$\Delta'_{0.5} = -\frac{L_1^2}{EI_0}(0.0625 \times 0.0489PL_1 - 0.013021 \times 0.0489PL_1 \times 0.281)$$

$$= -\frac{PL_1^3}{EI_0} 0.002877 \text{ in upward}$$

Therefore, the deflection of the beam at point 0.5 in Fig. 6-8 is

$$\Delta_{0.5} = \frac{PL_1^3}{EI_0}(0.020467 - 0.002877)$$

$$= \frac{PL_1^3}{EI_0} 0.017590 \text{ in downward} \qquad (ans.)$$

Example 6-8

Consider the same continuous beam shown in Example 6-7, but loaded with a uniformly distributed load (beam weight included), as shown in Fig. 6-9. Determine the deflection at the center of the spans AB and BC in terms of wL_1^4/EI_0. Let $L_1 : L_2 : L_3 = 1 : 1.333 : 1$.

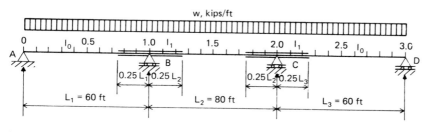

Fig. 6-9.

Solution

The fixed-end moments for a uniformly distributed load are

$$M_{AB} = 0.079wL_1^2 \text{ (Chart 36)} \qquad M_{BA} = 0.093wL_1^2 \text{ (Chart 34)}$$

$$M_{BC} = M_{CB} = 0.0885wL_2^2 = 0.157wL_1^2 \text{ (Chart 35)}$$

The moment-distribution information for Example 6-8 is given in the accompanying table.

Joint	A	B		C		D
Member	AB	BA	BC	CB	CD	DC
K	4.15	4.94	3.86	3.86	4.94	4.15
Modified K	—	3.586	1.749	1.749	3.586	—
Moment-distribution factor	1.00	0.672	0.328	0.328	0.672	1.00
Carryover factor	0.571	0.480	0.547	0.547	0.480	0.571
Fixed-end moment* balance	+7.90 −7.90	−9.30 −4.32	+15.73 −2.11	−15.73 +2.11	+9.30 +4.32	−7.90 +7.90
Carryover balance	−2.07 +2.07	−4.51 +2.26	+1.15 +1.10	−1.15 −1.10	+4.51 −2.26	+2.07 −2.07
Carryover balance	+1.08 −1.08	+1.18 −0.39	−0.60 −0.19	+0.60 +0.19	−1.18 +0.39	−1.08 +1.08
Carryover balance	−0.19 +0.19	−0.62 +0.35	+0.10 +0.17	−0.10 −0.17	+0.62 −0.35	+0.19 −0.19
Carryover balance	+0.17 −0.17	+0.11 −0.01	−0.09 −0.01	+0.09 +0.01	−0.11 +0.01	−0.17 +0.17
Total†	0	−15.25	+15.25	−15.25	+15.25	0

*Multiply figures by $wL_1/100$ to get fixed-end moments.
†$M = (wL_1^2/100) \times$ Total.

A. Compute the deflection at the center of span AB with the cover plates at the right end:

$$M_{AB} = 0 \qquad M_{BA} = -0.1525wL_1^2 \qquad \text{kip} \cdot \text{ft}$$

$$f_2(p') = 0.001058 \text{ (from Table 5-4)}$$

$$f_{BA}(p') = 0.013021 \text{ (from Table 5-6)}$$

The deflection for a simple-support, cover-plated beam is, by Eq. (5-10),

$$\Delta_{0.5} = \frac{wL_1^4}{EI_0}(0.013021 - 0.001058 \times 0.281)$$

$$= \frac{wL_1^4}{EI_0} 0.012724 \text{ in downward}$$

The reduction in deflection effected by the end moments is, by Eq. (5-14),

$$\Delta'_{0.5} = -\frac{L_1^2}{EI_0}(0.0625 \times 0.1525wL_1^2 - 0.013021 \times 0.1525wL_1^2 \times 0.281)$$

$$= -\frac{wL_1^4}{EI_0} 0.008973 \text{ in upward}$$

Therefore, the deflection of the beam at point 0.5 in Fig. 6-9 is

$$\Delta_{0.5} = \frac{wL_1^4}{EI_0}(0.012724 - 0.008977)$$

$$= \frac{wL_1^4}{EI_0} 0.003747 \text{ in downward} \qquad \text{(ans.)}$$

B. Compute the deflection at the center of the span BC with the cover plates at both ends:

$$M_{BC} = 0.1525wL_1^2 \quad \text{kip} \cdot \text{ft} \qquad M_{CB} = -0.1525wL_1^2 \quad \text{kip} \cdot \text{ft}$$

$$f_2(p) = f_2(p') = 0.001058 \text{ (from Table 5-4)}$$

$$f_{BC} = 0.01302 \quad \text{and} \quad f_{CB} = 0.002604 \text{ (from Table 5-6)}$$

The deflection for a simple-support, cover-plated beam is, by Eq. (5-10),

$$\Delta_{1.5} = \frac{wL_2^4}{EI_0}(0.013021 - 0.001058 \times 2 \times 0.281)$$

$$= \frac{wL_1^4}{EI_0} 0.039237 \text{ in downward}$$

The reduction in deflection effected by the end moments is, by Eq. (5-14),

$$\Delta'_{1.5} = -\frac{L_1^2}{EI_0}\Big\{0.0625 \times 2 \times 0.1525wL_1^2$$

$$- [(0.013021 + 0.002604) \times 2 \times 0.1525wL_2^2]0.281\Big\}$$

$$= -\frac{wL_1^4}{EI_0} 0.031494 \text{ in upward}$$

Therefore, the deflection of the beam at point 1.5 in Fig. 6-9 is

$$\Delta_{1.5} = \frac{wL_1^4}{EI_0}(0.039237 - 0.031494)$$

$$= \frac{wL_1^4}{EI_0} 0.007743 \text{ in downward} \qquad (ans.)$$

In the above examples, the charts were used to obtain the relative stiffnesses, carryover factors, and fixed-end moments. If equations are used, the results will be the same.

7
Deflection of Trusses

1 INTRODUCTION

When a truss is loaded, the joints of the truss will be deflected and the horizontal axis of the truss will become curved. Since sagging is not desirable, the usual practice is to raise the intermediate panel points above the horizontal line, so that when the truss is fully loaded, it will not deflect below the horizontal line at any point. Such an adjustment is known as cambering the truss. In short spans this may be done by the arbitrary rule of increasing each upper-chord member, the compression member, ⅛ in for each 10 ft of length and altering the diagonals to fit. In longer spans, that is, more than 300 ft, the deflection under dead load, or dead load plus live load, should be determined exactly and adjustments in the length of the members made accordingly. Several methods have been used for determining the deflection of trusses.

2 UNIT-LOAD METHOD

To find the deflections by the unit-load method, a unit load should be placed at the point in the direction that the deflection is to be found. Then the stresses u in each member due to this unit load can be determined. The required deflection is the sum of the products of ΔL and u for each member, where $\Delta L = SL/AE$. Thus, the deflection of the truss is

$$\Delta = \sum \frac{SuL}{AE} \qquad (7\text{-}1)$$

where A = cross-sectional area of member
E = modulus of elasticity of member
L = length of member
S = stress due to applied load
u = stress due to unit load applied at the point in the direction of the deflection to be computed

This is a working equation by which, at any point of a statically determinate truss, the deflection due to applied loads can be calculated.

Example 7-1

Determine the vertical deflections of all lower-chord points by the unit-load method. The complete data with regard to the truss dimensions, areas of members, and loading conditions are shown in Fig. 7-1(a). If the top-chord member were fabricated ³⁄₁₆ in longer, find the camber at the lower chord. Let $E = 30 \times 10^3$ kips/in².

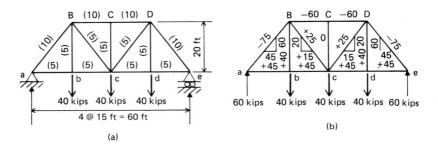

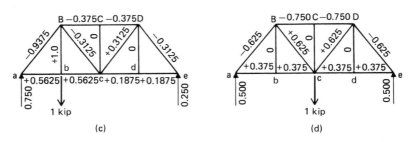

Fig. 7-1. (a) Numbers in parentheses are areas, in sq. in.; (b) values of S, in kips; (c) u_b for Δv of b; (d) u_c for Δv of c.

Solution

The solution of the stresses S, u_b, and u_c are shown in Fig. 7-1(b), (c), and (d), respectively. Stresses u_b and u_d are symmetrical about the centerline, so $u_b = u_d$. The deflections Δ_b, Δ_c, and Δ_d are obtained as shown in the table on the opposite page.

$$\Delta_b = +0.2657 \text{ in downward}$$
$$\Delta_c = +0.2910 \text{ in downward} \qquad (ans.)$$
$$\Delta_d = +0.2657 \text{ in downward}$$

Member	L/AE, in/kip	SL/AE, in	Su_bL/AE, in	Su_cL/AE, in
BC	0.0006	−0.036	+0.0135	+0.0270
CD	0.0006	−0.036	+0.0135	+0.0270
ab	0.0012	+0.054	+0.0304	+0.0202
bc	0.0012	+0.054	+0.0304	+0.0202
cd	0.0012	+0.054	+0.0101	+0.0202
de	0.0012	+0.054	+0.0101	+0.0202
Ba	0.0010	−0.075	+0.0703	+0.0469
Bc	0.0020	+0.050	−0.0156	+0.0312
Dc	0.0020	+0.050	+0.0156	+0.0312
De	0.0010	−0.075	+0.0234	+0.0469
Bb	0.0016	+0.064	+0.0640	0
Cc	0.0016	0	0	0
Dd	0.0016	+0.064	0	0
Σ	—	—	+0.2657	+0.2910

The plus sign indicates downward deflection.

When the top chord is increased by 3⁄16 in, that means $\Delta L = 0.1875$ in for members BC and CD. Then, from Fig. 7-1(c) and (d) and by $\Sigma\, u\Delta L$, we obtain

$$\Delta_b = 2 \times 0.1875 \times (-0.375) = -0.1406 \text{ in upward}$$

$$\Delta_c = 2 \times 0.1875 \times (-0.750) = -0.2812 \text{ in upward} \quad (ans.)$$

$$\Delta_d = 2 \times 0.1875 \times (-0.375) = -0.1406 \text{ in upward}$$

MAXIMUM DEFLECTION OF TRUSSES

In Example 7-1, the panel loads are given. In the case of a traffic load, the maximum panel load has to be computed in order to determine the maximum deflection of the truss. The maximum deflection of the truss is caused by the maximum panel load, which is a reaction from the floor beam. The value of the maximum panel load varies with the roadway width and the number of traffic lanes. The following example illustrates how to determine the maximum panel load on a truss.

Example 7-2

Find the maximum panel load for the highway bridge truss shown in Fig. 7-2. The traffic load is H20 loading. The roadway is 28 ft wide. The specifications cited in AASHTO's *Standard Specifications for Highway Bridges* are used.

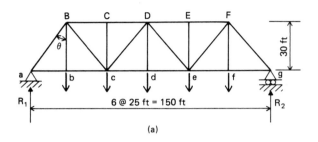

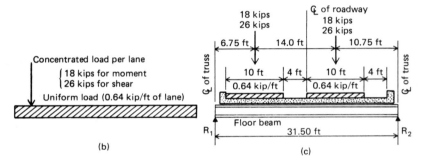

Fig. 7-2. Maximum panel loading on bridge truss. (a) Through Pratt truss; (b) H2O equivalent loading; (c) lane loading.

Solution

According to AASHTO specifications, the H20 equivalent loading is as shown in Fig. 7-2(b). Each traffic loading is 10 ft. The maximum reaction at the left-hand panel occurs when the traffic is placed at the position shown in Fig. 7-2(c). Thus.

$$R_{max} = \frac{18 \times 24.75 + 18 \times 10.75}{31.5} = 20.28 \text{ kips for moment}$$

$$R_{max} = \frac{26 \times 24.75 + 10.75 \times 10.75}{31.5} = 29.30 \text{ kips for shear} \quad (ans.)$$

$$R_{max} = \frac{0.64 \times 24.75 + 0.64 \times 10.75}{31.5} = 0.721 \text{ kip/ft of truss}$$

4 INFLUENCE LINES

Influence lines are important in determining the position of live load that will produce a maximum stress in truss members. Influence lines are also important in calculating the deflection of trusses. The deflection curve for the truss due to a change in length of a given member is the influence line for the stress of this member multiplied by ΔL. Thus, when a unit load is placed at a panel of the truss, that is the influence line for the vertical deflection at that point. In Example 7-1, the values of u_b, u_c, and u_d are the influence lines for the truss member stresses; when multiplied by ΔL, the deflection curve for the truss is obtained.

It is usual to use influence lines in the solution of stresses and deflections and to use the superposition method for solution of the combined problem. It can reduce the problem from a complicated case into a few basic and simple cases. The deflection of a truss due to a total force can be found by using the sum of the deflections due to each of the forces in combination. Also, it is more convenient to express influence data in the form of influence tables rather than in the form of influence curves in solving the deflections of trusses. Example 7-4 will illustrate this situation.

Example 7-3

For the bridge truss shown in Fig. 7-2(a), use influence lines to determine the maximum live load stresses and impact stresses for all web members and chord members. The AASHTO specifications are again used.

The maximum panel loads on the truss, obtained from Example 7-2, are

Concentrated load = 20.28 kips for moment
Concentrated load = 29.30 kips for shear
Uniform load = 0.721 kip/ft of truss

Solution

According to the AASHTO, the impact stress of a live load for highway bridges is determined by the formula

$$I = \frac{50}{L + 125} \qquad (7\text{-}2)$$

In this formula, I is the impact fraction whose maximum value is 30 percent of the live load stress, and L is the length in feet of the portion of the span which is loaded in order to produce the maximum live load stress for the member.

Influence lines for shear in panels ab, bc, and cd are shown in Fig. 7-3(b), (c), and (d), respectively. The maximum stress in a web member is equal to the maximum shear in its panel times sec θ, so the maximum live load stresses in web members are

$$S_{Ba} = -0.833(29.30 + \tfrac{1}{2} \times 0.721 \times 150) \times 1.30$$
$$= -90.3 \text{ kips } (C)$$

$$S_{Bc} = +0.666(29.30 + \tfrac{1}{2} \times 0.721 \times 120) \times 1.30$$
$$= +62.8 \text{ kips } (T) \quad (ans.)$$

$$S_{Cd} = -0.500(29.30 + \tfrac{1}{2} \times 0.721 \times 90) \times 1.30$$
$$= +40.1 \text{ kips } (T)$$

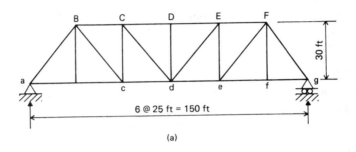

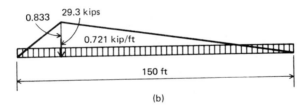

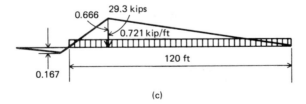

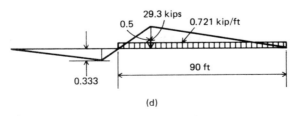

Fig. 7-3. (a) Through Pratt truss. Influence lines for shear in **(b)** panel *ab*; **(c)** panel *bc*; and **(d)** panel *cd*.

where C = compression and T = tension. The impact stresses in web members are

$$S_{Ba} = \frac{90.3 \times 50}{150 + 125} = 16.4 \text{ kips}$$

$$S_{Bc} = \frac{62.8 \times 50}{120 + 125} = 12.8 \text{ kips} \qquad (ans.)$$

$$S_{Cd} = \frac{40.1 \times 50}{90 + 125} = 9.3 \text{ kips}$$

The necessary influence lines for finding the maximum live load stresses in all chord members of the truss are shown in Fig. 7-4(b), (c), and (d),

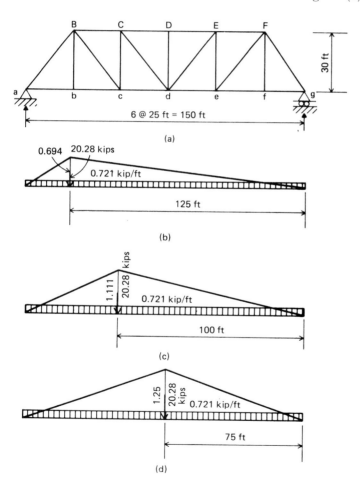

Fig. 7-4. (a) Through Pratt truss. Influence lines (b) for stress in ac; (c) for stress in cd and BC; (d) for stress in CD.

respectively. A peak in the influence line occurs under the center of moment for the chord member CE.

$$S_{BC} = -1.111(20.28 + \tfrac{1}{2} \times 0.721 \times 150) = -74.3 \text{ kips } (C)$$
$$S_{CD} = -1.250(20.28 + \tfrac{1}{2} \times 0.721 \times 150) = -92.9 \text{ kips } (C)$$
$$S_{ac} = +0.694(20.28 + \tfrac{1}{2} \times 0.721 \times 150) = +51.6 \text{ kips } (T)$$
$$S_{cd} = +1.111(20.28 + \tfrac{1}{2} \times 0.721 \times 150) = +74.3 \text{ kips } (T)$$

(ans.)

The impact fraction for chord members is constant for any one chord member of the truss since the entire span is loaded in all cases for the maximum chord stresses. The value of the impact fraction for chord members is $50/(150 + 125) = 0.18$. The maximum live load stresses and impact stresses for all web members and chord members are shown in Fig. 7-5.

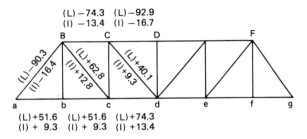

Fig. 7-5. Impact stresses and live load stresses from Example 7-3.

Example 7-4

Find the maximum vertical deflections of all lower-chord points in Example 7-3 by the unit-load method. Let $E = 30 \times 10^3$ kips/in^2. Use a concentrated load of 29.30 kips and a uniform load of 0.721 kip/ft of truss. Since the entire span is loaded in all cases, use $I = 0.18$ for all deflections.

Solution

Compute the deflection at each lower-chord point by placing a unit load at that point as shown in Fig. 7-6(b), (c), and (d). Then use the influence table on page 174 to compute the deflection at any chord point. Figure 7-6(e) gives the member stress Σu for the sum of the unit loads u_b, u_c, u_d, u_e, and u_f. The uniformly distributed load at each panel = $25 \times 0.721 = 18.025$ kips.

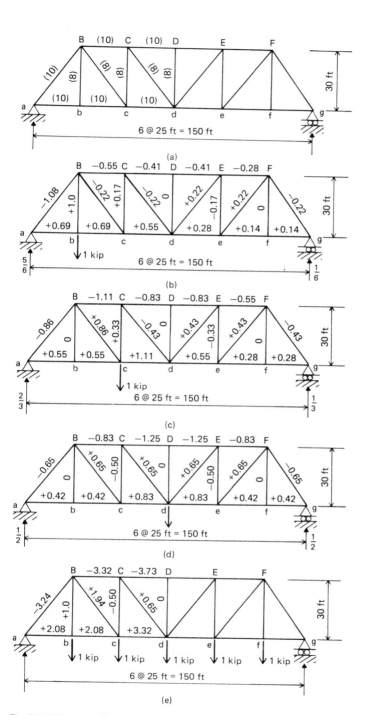

Fig. 7-6. (a) Through Pratt truss. Numbers in parentheses are areas in sq. in. (b) u_b for Δv of b; u_c for Δv of c; (d) u_d for Δv of d; (e) sum of unit load u_b, u_c, u_d, u_e, and u_f.

Member	L/AE, in/kip	Δ_b Concentrated unit load $u_b^2 L/AE$, in/kip	Δ_b Uniform unit load $\Sigma u\, u_b L/AE$, in/kip	Δ_c Concentrated unit load $u_c^2 L/AE$, in/kip	Δ_c Uniform unit load $\Sigma u\, u_c L/AE$, in/kip	Δ_d Concentrated unit load $u_d^2 L/AE$, in/kip	Δ_d Uniform unit load $\Sigma u\, u_d L/AE$, in/kip
BC	0.00100	+0.000302	+0.001826	+0.001232	+0.003685	+0.000689	+0.002756
CD	0.00100	+0.000168	+0.001529	+0.000689	+0.003096	+0.001562	+0.004662
DE	0.00100	+0.000168	+0.001529	+0.000689	+0.003096	+0.001562	+0.004662
EF	0.00100	+0.000078	+0.000930	+0.000302	+0.001826	+0.000689	+0.002756
ab	0.00100	+0.000476	+0.001435	+0.000302	+0.001144	+0.000176	+0.000874
bc	0.00100	+0.000476	+0.001435	+0.000302	+0.001144	+0.000176	+0.000874
cd	0.00100	+0.000302	+0.001826	+0.001232	+0.003685	+0.000689	+0.002789
de	0.00100	+0.000078	+0.000930	+0.000302	+0.001826	+0.000689	+0.002789
ef	0.00100	+0.000020	+0.000291	+0.000078	+0.000582	+0.000176	+0.000874
fg	0.00100	+0.000020	+0.000291	+0.000078	+0.000582	+0.000176	+0.000874
Ba	0.00156	+0.001820	+0.000459	+0.001154	+0.004347	+0.000659	+0.003285
Bc	0.00195	+0.000094	−0.000427	+0.001442	+0.003253	+0.000824	+0.002459
Cd	0.00195	+0.000094	−0.000279	+0.000360	−0.000545	+0.000824	+0.000824
Ed	0.00195	+0.000094	+0.000279	+0.000360	+0.000545	+0.000824	+0.000824
Fe	0.00195	+0.000094	+0.000427	+0.000360	+0.001627	+0.000824	+0.002459
Fg	0.00156	+0.000075	+0.001112	+0.000288	+0.002173	+0.000659	+0.003285
Bb	0.00120	+0.001200	+0.001200	0	0	0	0
Cc	0.00120	+0.000035	−0.000102	+0.000131	−0.000198	+0.000300	+0.000300
Dd	0.00120	0	0	0	0	0	0
Ee	0.00120	+0.000035	+0.000102	+0.000131	+0.000198	+0.000300	+0.000300
Fg	0.00120	0	0	0	0	0	0
Σ	—	+0.005629	+0.014793	+0.009432	+0.032066	+0.011798	+0.037646

Deflection of Trusses **175**

The maximum vertical deflections of all lower-chord points are

$$\Delta_b = (0.005629 \times 29.3 + 0.014793 \times 18.025)1.18$$
$$= 0.5090 \text{ in downward}$$

$$\Delta_c = (0.009432 \times 29.3 + 0.032066 \times 18.025)1.18$$
$$= 1.0081 \text{ in downward} \quad (ans.)$$

$$\Delta_d = (0.011798 \times 29.3 + 0.037646 \times 18.025)1.18$$
$$= 1.2086 \text{ in downward}$$

5 INDETERMINATE TRUSSES

The consistent deflection method has been used for solving statically indeterminate beams as is illustrated in Example 4-3. It can also be applied in solving indeterminate trusses. For example, let X be a redundant member. By conditions of geometry, $\Delta_x = X\delta_x$, where $\Delta_x = \Sigma_0^n (S'uL/AE)$ is the deflection due to applied load, and $\delta_x = \Sigma_0^{n+x} (u^2L/AE)$ is the deflection due to unit load. Thus, for solving the truss in a structure that is indeterminate in the first degree, the following equation should hold

$$X = -\frac{\sum_0^n S'uL/AE}{\sum_0^{n+x} u^2L/AE} \tag{7-3}$$

where A = cross-sectional area of the truss member
  L = length of truss member
  X = stress in redundant member
  S' = stress due to applied load, redundant X not included
  u = stress due to unit load acting along the line X, redundant X included

Example 7-5

The truss shown in Fig. 7-6(a) of Example 7-4 requires an additional support at point d. Determine all reactions if 100 kips is loaded at point b (see Fig. 7-7).

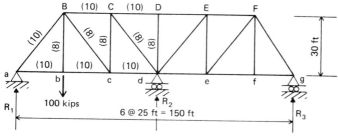

Fig. 7-7. Numbers in parentheses are areas in sq. in.

Solution

The truss has three supports and is an externally indeterminate truss to the first degree. Assume that R_2 is chosen as a redundant. From Example 7-4, Fig. 7-6(b) and (d), we obtain the values of S' and u_d and list the values in the accompanying table used for the solution of R_2. These values are used

Member	L/AE, in/kip	S', kips	u_d, 1 kip	$S'u_dL/AE$, in	u_d^2L/AE, in/kip
BC	0.00100	−55	−0.83	+0.045650	+0.000689
CD	0.00100	−41	−1.25	+0.051250	+0.001562
DE	0.00100	−41	−1.25	+0.051250	+0.001562
EF	0.00100	−28	−0.83	+0.023240	+0.000689
ab	0.00100	+69	+0.42	+0.028980	+0.000176
bc	0.00100	+69	+0.42	+0.028980	+0.000176
cd	0.00100	+55	+0.83	+0.045650	+0.000689
de	0.00100	+28	+0.83	+0.023240	+0.000689
ef	0.00100	+14	+0.42	+0.005880	+0.000176
fg	0.00100	+14	+0.42	+0.005880	+0.000176
Ba	0.00156	−108	−0.65	+0.109512	+0.000659
Bc	0.00195	−22	+0.65	−0.027885	+0.000824
Cd	0.00195	−22	+0.65	−0.027885	+0.000824
Ed	0.00195	+22	+0.65	+0.027885	+0.000824
Fe	0.00195	+22	+0.65	+0.027885	+0.000824
Fg	0.00156	−22	−0.65	+0.022308	+0.000659
Bb	0.00120	+100	0	0	0
Cc	0.00120	+17	−0.50	−0.010200	+0.000300
Dd	0.00120	0	0	0	0
Ee	0.00120	−17	−0.50	+0.010200	+0.000300
Ff	0.00120	0	0	0	0
Σ	—	—	—	+0.441820	+0.011798

in computing $S'u_dL/AE$ and u_d^2L/AE, from which the reaction of the redundant is found.

$$X = \frac{S'u_dL/AE}{u_d^2L/AE} = 37.45 \text{ kips}$$

$$R_1 = 64.61 \text{ kips}$$
$$R_2 = 37.45 \text{ kips} \qquad (ans.)$$
$$R_3 = -2.06 \text{ kips}$$

Example 7-6

A trussed beam consists of members as shown in Fig. 7-8. The beam carries a uniformly distributed load of 1.2 kips/ft. Find the stress in strut BD and in the rods AD and CD.

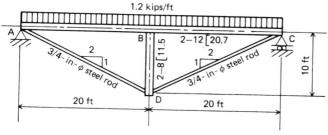

Fig. 7-8. Trussed beam.

Beam ABC: 2-12 [20.7 $A = 12.18$ in^2 $S = 43$ in^3 $I = 258$ in^4
Strut BD: 2-8 [11.5 $A = 6.76$ in^2
Rods AD and CD: ¾-in-φ $A = 1.502$ in^2
$E = 30 \times 10^3$ kips/in^2

Solution

Since the member ABC is subjected to both direct stress and bending, the whole structure can be considered as the sum of two structures, as shown in Fig. 7-9(a), (b), and (c). From these conditions, the deflection can be obtained.

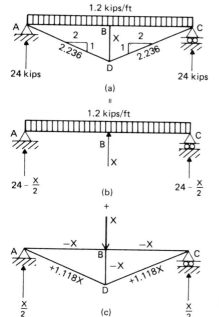

Fig. 7-9. Trussed beam stress analysis: (a) = (b) + (c).

The deflection Δ_B of the beam, see Fig. 7-9(b), by Eqs. 2-11 and 2-20, is

$$\Delta_B = \frac{1728 \times 40^3}{30 \times 10^3 \times 258}[0.013021 \times (1.2 \times 40) - 0.020833X]$$

$$= 8.9302 - 0.02976X$$

The deflection Δ'_B of the truss, see Fig. 7-9(c), by Eq. 7-1, is

$$\Delta'_B = \frac{12}{30 \times 10^3}\left[\frac{2X \times 1 \times 20}{12.18} + \frac{X \times 1 \times 10}{6.76} + \frac{2(1.118X)^2 \times 22.36}{1.502}\right]$$

$$= 0.0168X$$

Equating $\Delta_B = \Delta'_B$ and solving for X

$$X = 28.4 \text{ kips}$$

$$\text{Stress in strut } BD = \frac{28.4}{6.76} = 4.2 \text{ kips/in}^2 \; (C)$$

(ans.)

$$\text{Stress in rods } AD \text{ and } CD = \frac{1.118 \times 28.4}{1.502} = 21.1 \text{ kips/in}^2 \; (T)$$

Appendix

A
Beam Diagrams and Formulas for Various Static Loading Conditions

CASE 1 SIMPLE BEAM—CONCENTRATED LOAD AT CENTER*

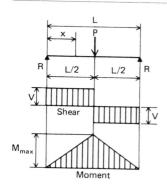

$$R = V = \frac{P}{2}$$

$$M_{max}\text{(at point of load)} = \frac{PL}{4}$$

$$M_x\left(\text{when } x < \frac{L}{2}\right) = \frac{Px}{2}$$

$$\Delta_{max}\text{(at point of load)} = \frac{PL^3}{48EI}$$

$$\Delta_x\left(\text{when } x < \frac{L}{2}\right) = \frac{Px}{48EI}(3L^2 - 4x^2)$$

CASE 2 SIMPLE BEAM—CONCENTRATED LOAD AT ANY POINT*

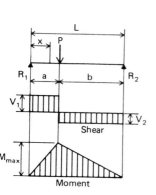

$$R_1 = V_1(\text{max when } a < b) = \frac{Pb}{L}$$

$$R_2 = V_2(\text{max when } a > b) = \frac{Pa}{L}$$

$$M_{max}\text{(at point of load)} = \frac{Pab}{L}$$

$$M_x(\text{when } x < a) = \frac{Pbx}{L}$$

$$\Delta_{max}\left(\text{at } x = \sqrt{\frac{a(a + 2b)}{3}} \text{ when } a > b\right)$$

$$= \frac{Pab(a + 2b)\sqrt{3a(a + 2b)}}{27EIL}$$

$$\Delta_a\text{(at point of load)} = \frac{Pa^2b^2}{3EIL}$$

$$\Delta_x(\text{when } x < a) = \frac{Pbx}{6EIL}(L^2 - b^2 - x^2)$$

Nomenclature: R = reaction, M = moment, V = shear, and Δ = deflection.
*From *Manual of Steel Construction*, 8th ed., American Institute of Steel Construction, 1980.

180 Appendix A

CASE 3 SIMPLE BEAM—TWO EQUAL CONCENTRATED LOADS SYMMETRICALLY PLACED*

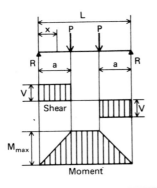

$$R = V = P$$
$$M_{max}(\text{between loads}) = Pa$$
$$M_x(\text{when } x < a) = Px$$
$$\Delta_{max}(\text{at center}) = \frac{Pa}{24EI}(3L^2 - 4a^2)$$
$$\Delta_x(\text{when } x < a) = \frac{Px}{6EI}(3La - 3a^2 - x^2)$$
$$\Delta_x[\text{when } x > a \text{ and } < (L - a)] = \frac{Pa}{6EI}(3Lx - 3x^2 - a^2)$$

CASE 4 SIMPLE BEAM—UNIFORMLY DISTRIBUTED LOAD*

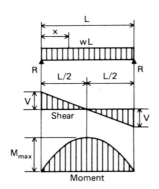

$$R = V = \frac{wL}{2}$$
$$V_x = w\left(\frac{L}{2} - x\right)$$
$$M_{max}(\text{at center}) = \frac{wL^2}{8}$$
$$M_x = \frac{wx}{2}(L - x)$$
$$\Delta_{max}(\text{at center}) = \frac{5wL^4}{384EI}$$
$$\Delta_x = \frac{wx}{24EI}(L^3 - 2Lx^2 + x^3)$$

Nomenclature: R = reaction, M = moment, V = shear, and Δ = deflection.
*From *Manual of Steel Construction*, 8th ed., American Institute of Steel Construction, 1980.

CASE 5 SIMPLE BEAM—UNIFORMLY DISTRIBUTED LOAD EXTENDING FROM ONE END*

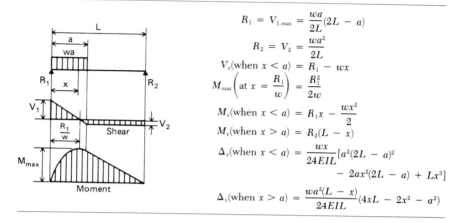

$$R_1 = V_{1,\max} = \frac{wa}{2L}(2L - a)$$

$$R_2 = V_2 = \frac{wa^2}{2L}$$

$$V_x(\text{when } x < a) = R_1 - wx$$

$$M_{\max}\left(\text{at } x = \frac{R_1}{w}\right) = \frac{R_1^2}{2w}$$

$$M_x(\text{when } x < a) = R_1 x - \frac{wx^2}{2}$$

$$M_x(\text{when } x > a) = R_2(L - x)$$

$$\Delta_x(\text{when } x < a) = \frac{wx}{24EIL}[a^2(2L - a)^2 - 2ax^2(2L - a) + Lx^3]$$

$$\Delta_x(\text{when } x > a) = \frac{wa^2(L - x)}{24EIL}(4xL - 2x^2 - a^2)$$

CASE 6 SIMPLE BEAM—LOAD INCREASING UNIFORMLY TO ONE END*

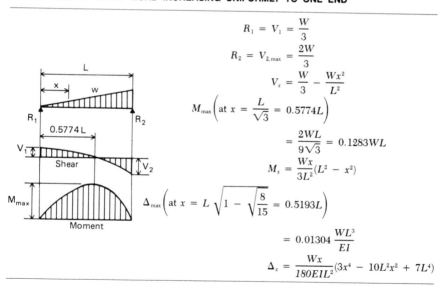

$$R_1 = V_1 = \frac{W}{3}$$

$$R_2 = V_{2,\max} = \frac{2W}{3}$$

$$V_x = \frac{W}{3} - \frac{Wx^2}{L^2}$$

$$M_{\max}\left(\text{at } x = \frac{L}{\sqrt{3}} = 0.5774L\right)$$

$$= \frac{2WL}{9\sqrt{3}} = 0.1283WL$$

$$M_x = \frac{Wx}{3L^2}(L^2 - x^2)$$

$$\Delta_{\max}\left(\text{at } x = L\sqrt{1 - \sqrt{\frac{8}{15}}} = 0.5193L\right)$$

$$= 0.01304\frac{WL^3}{EI}$$

$$\Delta_x = \frac{Wx}{180EIL^2}(3x^4 - 10L^2x^2 + 7L^4)$$

Nomenclature: R = reaction, M = moment, V = shear, and Δ = deflection.
*From *Manual of Steel Construction*, 8th ed., American Institute of Steel Construction, 1980.

CASE 7 SIMPLE BEAM—LOAD INCREASING UNIFORMLY TO CENTER*

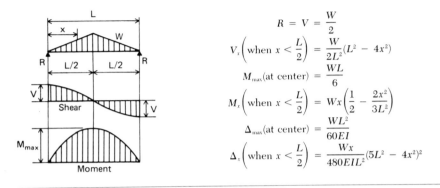

$$R = V = \frac{W}{2}$$

$$V_x \left(\text{when } x < \frac{L}{2} \right) = \frac{W}{2L^2}(L^2 - 4x^2)$$

$$M_{max}(\text{at center}) = \frac{WL}{6}$$

$$M_x \left(\text{when } x < \frac{L}{2} \right) = Wx \left(\frac{1}{2} - \frac{2x^2}{3L^2} \right)$$

$$\Delta_{max}(\text{at center}) = \frac{WL^3}{60EI}$$

$$\Delta_x \left(\text{when } x < \frac{L}{2} \right) = \frac{Wx}{480EIL^2}(5L^2 - 4x^2)^2$$

CASE 8 BEAM FIXED AT ONE END, SUPPORTED AT OTHER—CONCENTRATED LOAD AT CENTER*

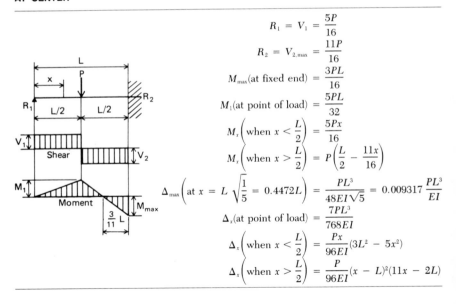

$$R_1 = V_1 = \frac{5P}{16}$$

$$R_2 = V_{2,max} = \frac{11P}{16}$$

$$M_{max}(\text{at fixed end}) = \frac{3PL}{16}$$

$$M_1(\text{at point of load}) = \frac{5PL}{32}$$

$$M_x \left(\text{when } x < \frac{L}{2} \right) = \frac{5Px}{16}$$

$$M_x \left(\text{when } x > \frac{L}{2} \right) = P \left(\frac{L}{2} - \frac{11x}{16} \right)$$

$$\Delta_{max} \left(\text{at } x = L\sqrt{\frac{1}{5}} = 0.4472L \right) = \frac{PL^3}{48EI\sqrt{5}} = 0.009317 \frac{PL^3}{EI}$$

$$\Delta_x(\text{at point of load}) = \frac{7PL^3}{768EI}$$

$$\Delta_x \left(\text{when } x < \frac{L}{2} \right) = \frac{Px}{96EI}(3L^2 - 5x^2)$$

$$\Delta_x \left(\text{when } x > \frac{L}{2} \right) = \frac{P}{96EI}(x - L)^2(11x - 2L)$$

Nomenclature: R = reaction, M = moment, V = shear, and Δ = deflection.
*From *Manual of Steel Construction*, 8th ed., American Institute of Steel Construction, 1980.

CASE 9 BEAM FIXED AT ONE END, SUPPORTED AT OTHER—CONCENTRATED LOAD AT ANY POINT*

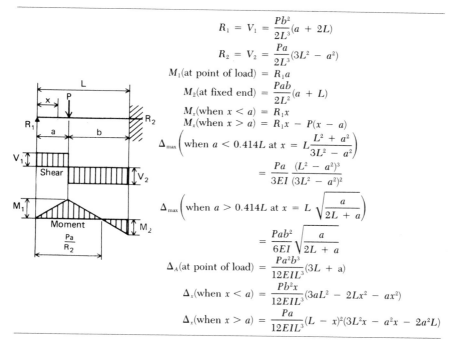

$$R_1 = V_1 = \frac{Pb^2}{2L^3}(a + 2L)$$

$$R_2 = V_2 = \frac{Pa}{2L^3}(3L^2 - a^2)$$

$$M_1(\text{at point of load}) = R_1 a$$

$$M_2(\text{at fixed end}) = \frac{Pab}{2L^2}(a + L)$$

$$M_x(\text{when } x < a) = R_1 x$$

$$M_x(\text{when } x > a) = R_1 x - P(x - a)$$

$$\Delta_{max}\left(\text{when } a < 0.414L \text{ at } x = L\frac{L^2 + a^2}{3L^2 - a^2}\right)$$

$$= \frac{Pa}{3EI} \frac{(L^2 - a^2)^3}{(3L^2 - a^2)^2}$$

$$\Delta_{max}\left(\text{when } a > 0.414L \text{ at } x = L\sqrt{\frac{a}{2L + a}}\right)$$

$$= \frac{Pab^2}{6EI}\sqrt{\frac{a}{2L + a}}$$

$$\Delta_A(\text{at point of load}) = \frac{Pa^2 b^3}{12EIL^3}(3L + a)$$

$$\Delta_x(\text{when } x < a) = \frac{Pb^2 x}{12EIL^3}(3aL^2 - 2Lx^2 - ax^2)$$

$$\Delta_x(\text{when } x > a) = \frac{Pa}{12EIL^3}(L - x)^2(3L^2 x - a^2 x - 2a^2 L)$$

CASE 10 BEAM FIXED AT ONE END, SUPPORTED AT OTHER—UNIFORMLY DISTRIBUTED LOAD*

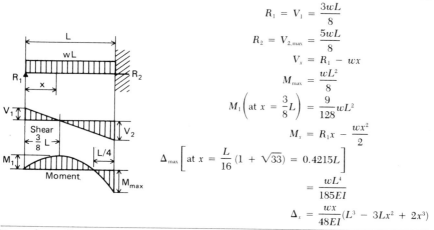

$$R_1 = V_1 = \frac{3wL}{8}$$

$$R_2 = V_{2,max} = \frac{5wL}{8}$$

$$V_x = R_1 - wx$$

$$M_{max} = \frac{wL^2}{8}$$

$$M_1\left(\text{at } x = \frac{3}{8}L\right) = \frac{9}{128}wL^2$$

$$M_x = R_1 x - \frac{wx^2}{2}$$

$$\Delta_{max}\left[\text{at } x = \frac{L}{16}(1 + \sqrt{33}) = 0.4215L\right]$$

$$= \frac{wL^4}{185EI}$$

$$\Delta_x = \frac{wx}{48EI}(L^3 - 3Lx^2 + 2x^3)$$

Nomenclature: R = reaction, M = moment, V = shear, and Δ = deflection.
*From *Manual of Steel Construction*, 8th ed., American Institute of Steel Construction, 1980.

CASE 11 BEAM FIXED AT ONE END, SUPPORTED AT THE OTHER—UNIFORMLY DISTRIBUTED LOAD EXTENDING FROM ONE END

$$R_1 = V_1 = \frac{wa}{8L^3}(8L^3 - 6aL^2 + a^3)$$

$$R_2 = V_2 = \frac{wa^2}{8L^3}(6L^2 - a^2)$$

$$M_{max}\left(\text{at } x = \frac{V_1}{w}\right) = \frac{V_1^2}{2w}$$

$$M_1(\text{at distance } a) = \frac{a}{2}(V_1 - V_2)$$

$$M_2(\text{at fixed end}) = \frac{wa^2}{8}\left(2 - \frac{a^2}{L^2}\right)$$

$$M_x(\text{when } x < a) = \frac{wax}{4L}\left(4L - 3a + \frac{a^3}{2L^2}\right) - \frac{1}{2}wx^2$$

$$M_x(\text{when } x > a) = \frac{wax}{4L}\left(4L - 3a + \frac{a^3}{2L^2}\right) - wa\left(x - \frac{1}{2}a\right)$$

$$\Delta_x(\text{when } x < a) = \frac{wx}{24EIL}\left[a^2(2L - a)^2 - 2a(2L - a)x^2 + Lx^3 - \frac{1}{2}\frac{a^2}{L^2}(2L^2 - a^2)(L^2 - x^2)\right]$$

$$\Delta_x(\text{when } x > a) = \frac{wa^2}{24EIL^2}\left[L(L - x)(4Lx - 2x^2 - a^2) - \frac{1}{2}\frac{x}{L}(2L^2 - a^2)(L^2 - x^2)\right]$$

CASE 12 BEAM FIXED AT ONE END, SUPPORTED AT THE OTHER—LOAD INCREASING UNIFORMLY TO THE FIXED END

$$R_1 = V_1 = \frac{W}{5}$$

$$R_2 = V_2 = \frac{4W}{5}$$

$$M_{max}(\text{at fixed end}) = -\frac{2WL}{15}$$

$$M_{max}(\text{at } x = 0.447L) = 0.06WL$$

$$M_x = W\left(\frac{1}{5}x - \frac{1}{3}\frac{x^3}{L^2}\right)$$

$$\Delta_{max}\left(\text{at } x = \sqrt{\frac{1}{5}}\right) = 0.00477\frac{WL^3}{EI}$$

$$\Delta_x = \frac{W}{60EIL}\left(2Lx^3 - L^3x - \frac{x^5}{L}\right)$$

Nomenclature: R = reaction, M = moment, V = shear, and Δ = deflection.

CASE 13 BEAM FIXED AT BOTH ENDS—CONCENTRATED LOAD AT CENTER*

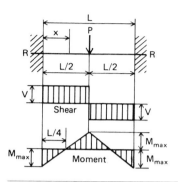

$$R = V = \frac{P}{2}$$

$$M_{max}(\text{at center and ends}) = \frac{PL}{8}$$

$$M_x\left(\text{when } x < \frac{L}{2}\right) = \frac{P}{8}(4x - L)$$

$$\Delta_{max}(\text{at center}) = \frac{PL^3}{192EI}$$

$$\Delta_x\left(\text{when } x < \frac{L}{2}\right) = \frac{Px^2}{48EI}(3L - 4x)$$

CASE 14 BEAM FIXED AT BOTH ENDS—CONCENTRATED LOAD AT ANY POINT*

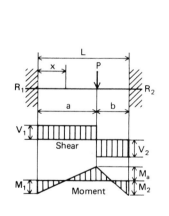

$$R_1 = V_1(\text{max when } a < b) = \frac{Pb^2}{L^3}(3a + b)$$

$$R_2 = V_2(\text{max when } a > b) = \frac{Pa^2}{L^3}(a + 3b)$$

$$M_1(\text{max when } a < b) = \frac{Pab^2}{L^2}$$

$$M_2(\text{max when } a > b) = \frac{Pa^2b}{L^2}$$

$$M_a(\text{at point of load}) = \frac{2Pa^2b^2}{L^3}$$

$$M_x(\text{when } x < a) = R_1x - \frac{Pab^2}{L^2}$$

$$\Delta_{max}\left(\text{when } a > b \text{ at } x = \frac{2aL}{3a + b}\right)$$

$$= \frac{2Pa^3b^2}{3EI(3a + b)^2}$$

$$\Delta_a(\text{at point of load}) = \frac{Pa^3b^3}{3EIL^3}$$

$$\Delta_x(\text{when } x < a) = \frac{Pb^2x^2}{6EIL^3}(3aL - 3ax - bx)$$

Nomenclature: R = reaction, M = moment, V = shear, and Δ = deflection.
*From *Manual of Steel Construction*, 8th ed., American Institute of Steel Construction, 1980.

CASE 15 BEAM FIXED AT BOTH ENDS—TWO EQUAL CONCENTRATED LOADS SYMMETRICALLY PLACED

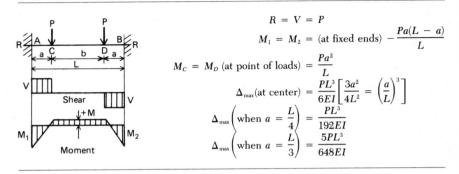

$$R = V = P$$
$$M_1 = M_2 = \text{(at fixed ends)} - \frac{Pa(L-a)}{L}$$
$$M_C = M_D \text{ (at point of loads)} = \frac{Pa^2}{L}$$
$$\Delta_{max}\text{(at center)} = \frac{PL^3}{6EI}\left[\frac{3a^2}{4L^2} - \left(\frac{a}{L}\right)^3\right]$$
$$\Delta_{max}\left(\text{when } a = \frac{L}{4}\right) = \frac{PL^3}{192EI}$$
$$\Delta_{max}\left(\text{when } a = \frac{L}{3}\right) = \frac{5PL^3}{648EI}$$

CASE 16 BEAM FIXED AT BOTH ENDS—THREE EQUAL CONCENTRATED LOADS SYMMETRICALLY PLACED

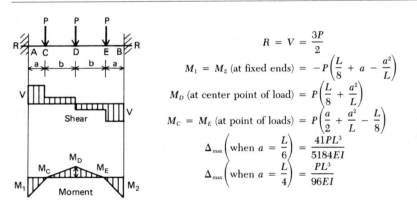

$$R = V = \frac{3P}{2}$$
$$M_1 = M_2 \text{ (at fixed ends)} = -P\left(\frac{L}{8} + a - \frac{a^2}{L}\right)$$
$$M_D \text{ (at center point of load)} = P\left(\frac{L}{8} + \frac{a^2}{L}\right)$$
$$M_C = M_E \text{ (at point of loads)} = P\left(\frac{a}{2} + \frac{a^2}{L} - \frac{L}{8}\right)$$
$$\Delta_{max}\left(\text{when } a = \frac{L}{6}\right) = \frac{41PL^3}{5184EI}$$
$$\Delta_{max}\left(\text{when } a = \frac{L}{4}\right) = \frac{PL^3}{96EI}$$

Nomenclature: R = reaction, M = moment, V = shear, and Δ = deflection.

CASE 17 BEAM FIXED AT BOTH ENDS—UNIFORMLY DISTRIBUTED LOAD*

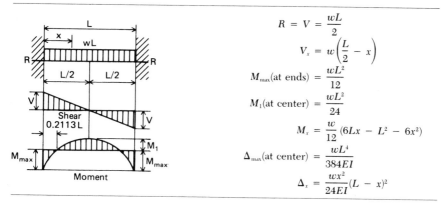

$$R = V = \frac{wL}{2}$$

$$V_x = w\left(\frac{L}{2} - x\right)$$

$$M_{max}(\text{at ends}) = \frac{wL^2}{12}$$

$$M_1(\text{at center}) = \frac{wL^2}{24}$$

$$M_x = \frac{w}{12}(6Lx - L^2 - 6x^2)$$

$$\Delta_{max}(\text{at center}) = \frac{wL^4}{384EI}$$

$$\Delta_x = \frac{wx^2}{24EI}(L - x)^2$$

CASE 18 BEAM FIXED AT BOTH ENDS—UNIFORMLY DISTRIBUTED LOAD EXTENDING FROM ONE END

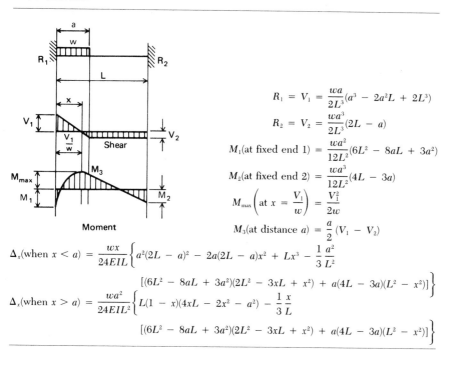

$$R_1 = V_1 = \frac{wa}{2L^3}(a^3 - 2a^2L + 2L^3)$$

$$R_2 = V_2 = \frac{wa^3}{2L^3}(2L - a)$$

$$M_1(\text{at fixed end 1}) = \frac{wa^2}{12L^2}(6L^2 - 8aL + 3a^2)$$

$$M_2(\text{at fixed end 2}) = \frac{wa^3}{12L^2}(4L - 3a)$$

$$M_{max}\left(\text{at } x = \frac{V_1}{w}\right) = \frac{V_1^2}{2w}$$

$$M_3(\text{at distance } a) = \frac{a}{2}(V_1 - V_2)$$

$$\Delta_x(\text{when } x < a) = \frac{wx}{24EIL}\left\{a^2(2L - a)^2 - 2a(2L - a)x^2 + Lx^3 - \frac{1}{3}\frac{a^2}{L^2}\right.$$

$$\left.[(6L^2 - 8aL + 3a^2)(2L^2 - 3xL + x^2) + a(4L - 3a)(L^2 - x^2)]\right\}$$

$$\Delta_x(\text{when } x > a) = \frac{wa^2}{24EIL^2}\left\{L(1 - x)(4xL - 2x^2 - a^2) - \frac{1}{3}\frac{x}{L}\right.$$

$$\left.[(6L^2 - 8aL + 3a^2)(2L^2 - 3xL + x^2) + a(4L - 3a)(L^2 - x^2)]\right\}$$

Nomenclature: R = reaction, M = moment, V = shear, and Δ = deflection.
*From *Manual of Steel Construction*, 8th ed., American Institute of Steel Construction, 1980.

CASE 19 BEAM FIXED AT BOTH ENDS—LOAD INCREASING UNIFORMLY TO THE RIGHT END

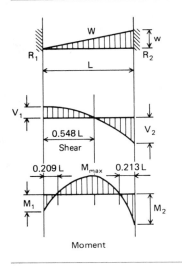

$$R_1 = V_1 = \frac{3W}{10}$$

$$R_2 = V_2 = \frac{7W}{10}$$

$$M_1 \text{ (at fixed end 1)} = \frac{WL}{15}$$

$$M_2 \text{ (at fixed end 2)} = -\frac{WL}{10}$$

$$M_{max}(\text{at } 0.548L) = 0.043WL$$

$$\Delta_{max}(\text{at } 0.525L) = \frac{WL^3}{382EI}$$

CASE 20 BEAM FIXED AT BOTH ENDS—LOAD INCREASING UNIFORMLY TO THE CENTER

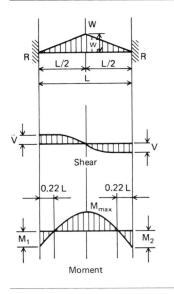

$$R = V = \frac{W}{2}$$

$$M_1 = M_2 \text{ (at fixed ends)} = -\frac{5WL}{48}$$

$$M_{max}(\text{at center}) = \frac{WL}{16}$$

$$\Delta_{max}(\text{at center}) = \frac{1.4WL^3}{384EI}$$

Nomenclature: R = reaction, M = moment, V = shear, and Δ = deflection.

CASE 21 CANTILEVER BEAM—CONCENTRATED LOAD AT ANY POINT*

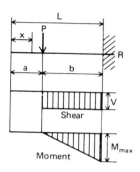

$$R = V = P$$
$$M_{max}(\text{at fixed end}) = Pb$$
$$M_x(\text{when } x > a) = P(x - a)$$
$$\Delta_{max}(\text{at free end}) = \frac{Pb^2}{6EI}(3L - b)$$
$$\Delta_a(\text{at point of load}) = \frac{Pb^3}{3EI}$$
$$\Delta_x(\text{when } x < a) = \frac{Pb^2}{6EI}(3L - 3x - b)$$
$$\Delta_x(\text{when } x > a) = \frac{P(L - x)^2}{6EI}(3b - L + x)$$

CASE 22 CANTILEVER BEAM—CONCENTRATED LOAD AT FREE END*

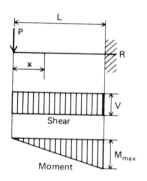

$$R = V = P$$
$$M_{max}(\text{at fixed end}) = PL$$
$$M_x = Px$$
$$\Delta_{max}(\text{at free end}) = \frac{PL^3}{3EI}$$
$$\Delta_x = \frac{P}{6EI}(2L^3 - 3L^2x + x^3)$$

Nomenclature: R = reaction, M = moment, V = shear, and Δ = deflection.
*From *Manual of Steel Construction*, 8th ed., American Institute of Steel Construction, 1980.

CASE 23 CANTILEVER BEAM—UNIFORMLY DISTRIBUTED LOAD*

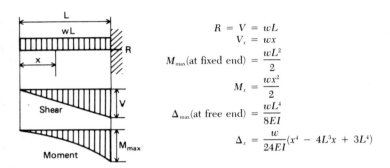

$$R = V = wL$$
$$V_x = wx$$
$$M_{max}\text{(at fixed end)} = \frac{wL^2}{2}$$
$$M_x = \frac{wx^2}{2}$$
$$\Delta_{max}\text{(at free end)} = \frac{wL^4}{8EI}$$
$$\Delta_x = \frac{w}{24EI}(x^4 - 4L^3x + 3L^4)$$

CASE 24 CANTILEVER BEAM—UNIFORMLY DISTRIBUTED LOAD EXTENDING TO THE FIXED END

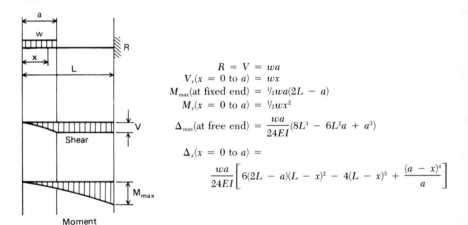

$$R = V = wa$$
$$V_x(x = 0 \text{ to } a) = wx$$
$$M_{max}\text{(at fixed end)} = \tfrac{1}{2}wa(2L - a)$$
$$M_x(x = 0 \text{ to } a) = \tfrac{1}{2}wx^2$$
$$\Delta_{max}\text{(at free end)} = \frac{wa}{24EI}(8L^3 - 6L^2a + a^3)$$

$$\Delta_x(x = 0 \text{ to } a) =$$
$$\frac{wa}{24EI}\left[6(2L - a)(L - x)^2 - 4(L - x)^3 + \frac{(a - x)^4}{a}\right]$$

Nomenclature: R = reaction, M = moment, V = shear, and Δ = deflection.
*From *Manual of Steel Construction*, 8th ed., American Institute of Steel Construction, 1980.

CASE 25 CANTILEVER BEAM—UNIFORM LOAD PARTIALLY DISTRIBUTED

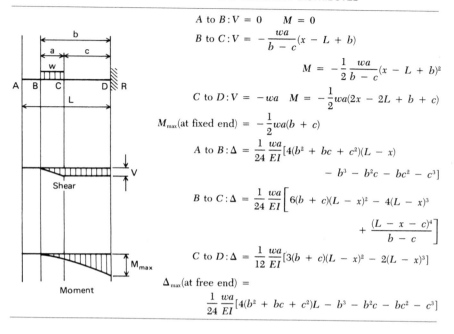

A to $B: V = 0 \quad M = 0$

B to $C: V = -\dfrac{wa}{b-c}(x - L + b)$

$\qquad M = -\dfrac{1}{2}\dfrac{wa}{b-c}(x - L + b)^2$

C to $D: V = -wa \quad M = -\dfrac{1}{2}wa(2x - 2L + b + c)$

$M_{\max}$(at fixed end) $= -\dfrac{1}{2}wa(b + c)$

A to $B: \Delta = \dfrac{1}{24}\dfrac{wa}{EI}[4(b^2 + bc + c^2)(L - x)$
$\qquad\qquad - b^3 - b^2c - bc^2 - c^3]$

B to $C: \Delta = \dfrac{1}{24}\dfrac{wa}{EI}\left[6(b + c)(L - x)^2 - 4(L - x)^3 + \dfrac{(L - x - c)^4}{b - c}\right]$

C to $D: \Delta = \dfrac{1}{12}\dfrac{wa}{EI}[3(b + c)(L - x)^2 - 2(L - x)^3]$

$\Delta_{\max}$(at free end) $=$
$\dfrac{1}{24}\dfrac{wa}{EI}[4(b^2 + bc + c^2)L - b^3 - b^2c - bc^2 - c^3]$

CASE 26 CANTILEVER BEAM—LOAD INCREASING UNIFORMLY TO FIXED END*

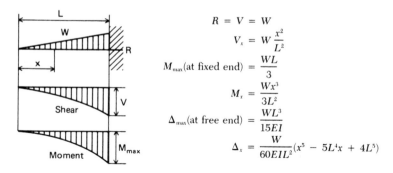

$R = V = W$

$V_x = W\dfrac{x^2}{L^2}$

$M_{\max}$(at fixed end) $= \dfrac{WL}{3}$

$M_x = \dfrac{Wx^3}{3L^2}$

$\Delta_{\max}$(at free end) $= \dfrac{WL^3}{15EI}$

$\Delta_x = \dfrac{W}{60EIL^2}(x^5 - 5L^4x + 4L^5)$

Nomenclature: R = reaction, M = moment, V = shear, and Δ = deflection.
*From *Manual of Steel Construction*, 8th ed., American Institute of Steel Construction, 1980.

CASE 27 BEAM FIXED AT ONE END, FREE TO DEFLECT VERTICALLY BUT NOT ROTATE AT OTHER—CONCENTRATED LOAD AT DEFLECTED END*

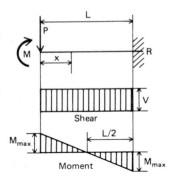

$$R = V = P$$
$$M_{max}(\text{at both ends}) = \frac{PL}{2}$$
$$M_x = P\left(\frac{L}{2} - x\right)$$
$$\Delta_{max}(\text{at deflected end}) = \frac{PL^3}{12EI}$$
$$\Delta_x = \frac{P(L-x)^2}{12EI}(L + 2x)$$

CASE 28 BEAM FIXED AT ONE END, FREE TO DEFLECT VERTICALLY BUT NOT ROTATE AT OTHER—UNIFORMLY DISTRIBUTED LOAD*

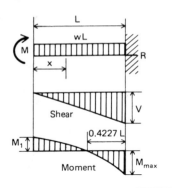

$$R = V = wL$$
$$V_x = wx$$
$$M_{max}(\text{at fixed end}) = \frac{wL^2}{3}$$
$$M_1(\text{at deflected end}) = \frac{wL^2}{6}$$
$$M_x = \frac{w}{6}(L^2 - 3x^2)$$
$$\Delta_{max}(\text{at deflected end}) = \frac{wL^4}{24EI}$$
$$\Delta_x = \frac{w(L^2 - x^2)^2}{24EI}$$

Nomenclature: R = reaction, M = moment, V = shear, and Δ = deflection.
*From *Manual of Steel Construction*, 8th ed., American Institute of Steel Construction, 1980.

CASE 29 BEAM OVERHANGING ONE SUPPORT—CONCENTRATED LOAD AT ANY POINT BETWEEN SUPPORTS*

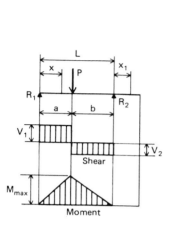

$$R_1 = V_1(\text{max when } a < b) = \frac{Pb}{L}$$

$$R_2 = V_2(\text{max when } a > b) = \frac{Pa}{L}$$

$$M_{\max}(\text{at point of load}) = \frac{Pab}{L}$$

$$M_x(\text{when } x < a) = \frac{Pbx}{L}$$

$$\Delta_{\max}\left(\text{at } x = \sqrt{\frac{a(a + 2b)}{3}} \text{ when } a > b\right)$$

$$= \frac{Pab(a + 2b)\sqrt{3a(a + 2b)}}{27EIL}$$

$$\Delta_a(\text{at point of load}) = \frac{Pa^2b^2}{3EIL}$$

$$\Delta_x(\text{when } x < a) = \frac{Pbx}{6EIL}(L^2 - b^2 - x^2)$$

$$\Delta_x(\text{when } x > a) = \frac{Pa(L - x)}{6EIL}(2Lx - x^2 - a^2)$$

$$\Delta_{x_1} = \frac{Pabx_1}{6EIL}(L + a)$$

CASE 30 BEAM OVERHANGING ONE SUPPORT—CONCENTRATED LOAD AT END OF OVERHANG*

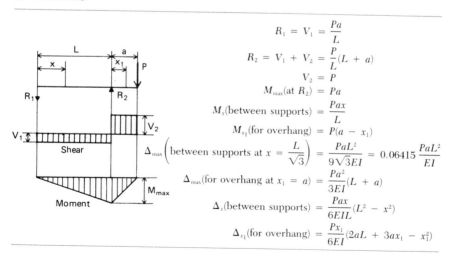

$$R_1 = V_1 = \frac{Pa}{L}$$

$$R_2 = V_1 + V_2 = \frac{P}{L}(L + a)$$

$$V_2 = P$$

$$M_{\max}(\text{at } R_2) = Pa$$

$$M_x(\text{between supports}) = \frac{Pax}{L}$$

$$M_{x_1}(\text{for overhang}) = P(a - x_1)$$

$$\Delta_{\max}\left(\text{between supports at } x = \frac{L}{\sqrt{3}}\right) = \frac{PaL^2}{9\sqrt{3}EI} = 0.06415\frac{PaL^2}{EI}$$

$$\Delta_{\max}(\text{for overhang at } x_1 = a) = \frac{Pa^2}{3EI}(L + a)$$

$$\Delta_x(\text{between supports}) = \frac{Pax}{6EIL}(L^2 - x^2)$$

$$\Delta_{x_1}(\text{for overhang}) = \frac{Px_1}{6EI}(2aL + 3ax_1 - x_1^2)$$

Nomenclature: R = reaction, M = moment, V = shear, and Δ = deflection.
*From *Manual of Steel Construction*, 8th ed., American Institute of Steel Construction, 1980.

CASE 31 BEAM OVERHANGING ONE SUPPORT—UNIFORMLY DISTRIBUTED LOAD*

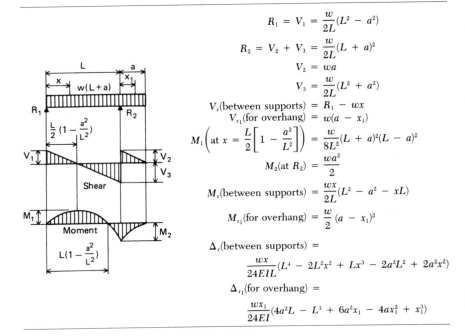

$$R_1 = V_1 = \frac{w}{2L}(L^2 - a^2)$$

$$R_2 = V_2 + V_3 = \frac{w}{2L}(L + a)^2$$

$$V_2 = wa$$

$$V_3 = \frac{w}{2L}(L^2 + a^2)$$

$$V_x(\text{between supports}) = R_1 - wx$$

$$V_{x_1}(\text{for overhang}) = w(a - x_1)$$

$$M_1\left(\text{at } x = \frac{L}{2}\left[1 - \frac{a^2}{L^2}\right]\right) = \frac{w}{8L^2}(L + a)^2(L - a)^2$$

$$M_2(\text{at } R_2) = \frac{wa^2}{2}$$

$$M_x(\text{between supports}) = \frac{wx}{2L}(L^2 - a^2 - xL)$$

$$M_{x_1}(\text{for overhang}) = \frac{w}{2}(a - x_1)^2$$

$$\Delta_x(\text{between supports}) =$$
$$\frac{wx}{24EIL}(L^4 - 2L^2x^2 + Lx^3 - 2a^2L^2 + 2a^2x^2)$$

$$\Delta_{x_1}(\text{for overhang}) =$$
$$\frac{wx_1}{24EI}(4a^2L - L^3 + 6a^2x_1 - 4ax_1^2 + x_1^3)$$

CASE 32 BEAM OVERHANGING ONE SUPPORT—UNIFORMLY DISTRIBUTED LOAD BETWEEN SUPPORTS*

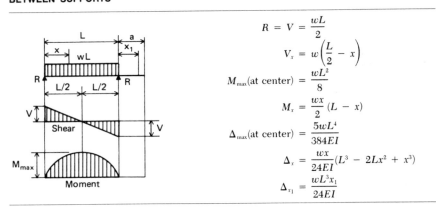

$$R = V = \frac{wL}{2}$$

$$V_x = w\left(\frac{L}{2} - x\right)$$

$$M_{\max}(\text{at center}) = \frac{wL^2}{8}$$

$$M_x = \frac{wx}{2}(L - x)$$

$$\Delta_{\max}(\text{at center}) = \frac{5wL^4}{384EI}$$

$$\Delta_x = \frac{wx}{24EI}(L^3 - 2Lx^2 + x^3)$$

$$\Delta_{x_1} = \frac{wL^3x_1}{24EI}$$

Nomenclature: R = reaction, M = moment, V = shear, and Δ = deflection.
*From *Manual of Steel Construction*, 8th ed., American Institute of Steel Construction, 1980.

CASE 33 BEAM OVERHANGING ONE SUPPORT—UNIFORMLY DISTRIBUTED LOAD ON OVERHANG*

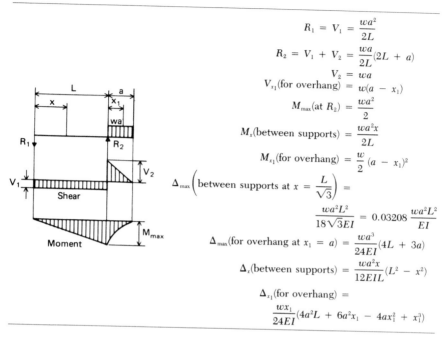

$$R_1 = V_1 = \frac{wa^2}{2L}$$

$$R_2 = V_1 + V_2 = \frac{wa}{2L}(2L + a)$$

$$V_2 = wa$$

$$V_{x_1}\text{(for overhang)} = w(a - x_1)$$

$$M_{\max}\text{(at } R_2\text{)} = \frac{wa^2}{2}$$

$$M_x\text{(between supports)} = \frac{wa^2 x}{2L}$$

$$M_{x_1}\text{(for overhang)} = \frac{w}{2}(a - x_1)^2$$

$$\Delta_{\max}\left(\text{between supports at } x = \frac{L}{\sqrt{3}}\right) =$$

$$\frac{wa^2 L^2}{18\sqrt{3}EI} = 0.03208 \frac{wa^2 L^2}{EI}$$

$$\Delta_{\max}\text{(for overhang at } x_1 = a\text{)} = \frac{wa^3}{24EI}(4L + 3a)$$

$$\Delta_x\text{(between supports)} = \frac{wa^2 x}{12EIL}(L^2 - x^2)$$

$$\Delta_{x_1}\text{(for overhang)} =$$

$$\frac{wx_1}{24EI}(4a^2 L + 6a^2 x_1 - 4ax_1^2 + x_1^3)$$

CASE 34 BEAM—CONCENTRATED LOAD AT CENTER AND VARIABLE END MOMENTS*

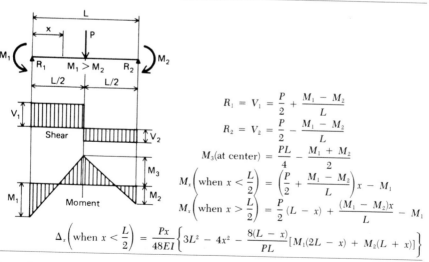

$$R_1 = V_1 = \frac{P}{2} + \frac{M_1 - M_2}{L}$$

$$R_2 = V_2 = \frac{P}{2} - \frac{M_1 - M_2}{L}$$

$$M_3\text{(at center)} = \frac{PL}{4} - \frac{M_1 + M_2}{2}$$

$$M_x\left(\text{when } x < \frac{L}{2}\right) = \left(\frac{P}{2} + \frac{M_1 - M_2}{L}\right)x - M_1$$

$$M_x\left(\text{when } x > \frac{L}{2}\right) = \frac{P}{2}(L - x) + \frac{(M_1 - M_2)x}{L} - M_1$$

$$\Delta_x\left(\text{when } x < \frac{L}{2}\right) = \frac{Px}{48EI}\left\{3L^2 - 4x^2 - \frac{8(L - x)}{PL}[M_1(2L - x) + M_2(L + x)]\right\}$$

Nomenclature: R = reaction, M = moment, V = shear, and Δ = deflection.
*From *Manual of Steel Construction*, 8th ed., American Institute of Steel Construction, 1980.

CASE 35 BEAM—UNIFORMLY DISTRIBUTED LOAD AND VARIABLE END MOMENTS*

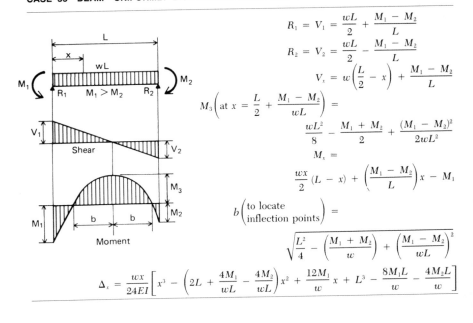

$$R_1 = V_1 = \frac{wL}{2} + \frac{M_1 - M_2}{L}$$

$$R_2 = V_2 = \frac{wL}{2} - \frac{M_1 - M_2}{L}$$

$$V_x = w\left(\frac{L}{2} - x\right) + \frac{M_1 - M_2}{L}$$

$$M_3 \left(\text{at } x = \frac{L}{2} + \frac{M_1 - M_2}{wL}\right) =$$

$$\frac{wL^2}{8} - \frac{M_1 + M_2}{2} + \frac{(M_1 - M_2)^2}{2wL^2}$$

$$M_x = \frac{wx}{2}(L - x) + \left(\frac{M_1 - M_2}{L}\right)x - M_1$$

$$b\binom{\text{to locate}}{\text{inflection points}} =$$

$$\sqrt{\frac{L^2}{4} - \left(\frac{M_1 + M_2}{w}\right) + \left(\frac{M_1 - M_2}{wL}\right)^2}$$

$$\Delta_x = \frac{wx}{24EI}\left[x^3 - \left(2L + \frac{4M_1}{wL} - \frac{4M_2}{wL}\right)x^2 + \frac{12M_1}{w}x + L^3 - \frac{8M_1 L}{w} - \frac{4M_2 L}{w}\right]$$

CASE 36 CONTINUOUS BEAM OF TWO EQUAL SPANS—CONCENTRATED LOAD AT CENTER OF ONE SPAN*

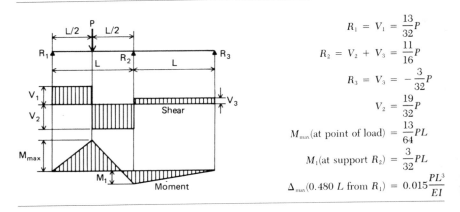

$$R_1 = V_1 = \frac{13}{32}P$$

$$R_2 = V_2 + V_3 = \frac{11}{16}P$$

$$R_3 = V_3 = -\frac{3}{32}P$$

$$V_2 = \frac{19}{32}P$$

$$M_{max}(\text{at point of load}) = \frac{13}{64}PL$$

$$M_1(\text{at support } R_2) = \frac{3}{32}PL$$

$$\Delta_{max}(0.480\ L \text{ from } R_1) = 0.015\frac{PL^3}{EI}$$

Nomenclature: R = reaction, M = moment, V = shear, and Δ = deflection.
*From *Manual of Steel Construction*, 8th ed., American Institute of Steel Construction, 1980.

CASE 37 CONTINUOUS BEAM OF TWO EQUAL SPANS—CONCENTRATED LOAD AT ANY POINT*

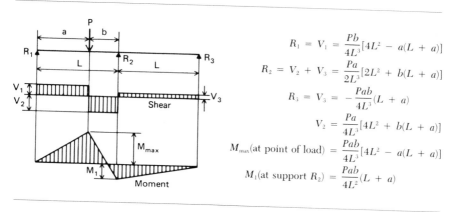

$$R_1 = V_1 = \frac{Pb}{4L^3}[4L^2 - a(L + a)]$$

$$R_2 = V_2 + V_3 = \frac{Pa}{2L^3}[2L^2 + b(L + a)]$$

$$R_3 = V_3 = -\frac{Pab}{4L^3}(L + a)$$

$$V_2 = \frac{Pa}{4L^3}[4L^2 + b(L + a)]$$

$$M_{max}(\text{at point of load}) = \frac{Pab}{4L^3}[4L^2 - a(L + a)]$$

$$M_1(\text{at support } R_2) = \frac{Pab}{4L^2}(L + a)$$

CASE 38 CONTINUOUS BEAM OF TWO EQUAL SPANS—UNIFORM LOAD ON ONE SPAN*

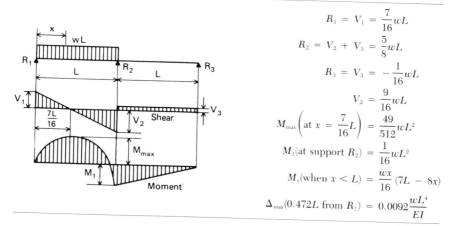

$$R_1 = V_1 = \frac{7}{16}wL$$

$$R_2 = V_2 + V_3 = \frac{5}{8}wL$$

$$R_3 = V_3 = -\frac{1}{16}wL$$

$$V_2 = \frac{9}{16}wL$$

$$M_{max}\left(\text{at } x = \frac{7}{16}L\right) = \frac{49}{512}wL^2$$

$$M_1(\text{at support } R_2) = \frac{1}{16}wL^2$$

$$M_x(\text{when } x < L) = \frac{wx}{16}(7L - 8x)$$

$$\Delta_{max}(0.472L \text{ from } R_1) = 0.0092\frac{wL^4}{EI}$$

Nomenclature: R = reaction, M = moment, V = shear, and Δ = deflection.
*From *Manual of Steel Construction*, 8th ed., American Institute of Steel Construction, 1980.

198 Appendix A

The values given in the formulas of the following cases do not include impact, which varies according to the requirements of each case.

CASE 39 SIMPLE BEAM—ONE CONCENTRATED, MOVING LOAD*

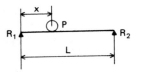

$$R_{1,\max} = V_{1,\max}(\text{at } x = o) = P$$

$$M_{\max}\left(\text{at point of load, when } x = \frac{L}{2}\right) = \frac{PL}{4}$$

CASE 40 SIMPLE BEAM—TWO EQUAL CONCENTRATED, MOVING LOADS*

$$R_{1,\max} = V_{1,\max}(\text{at } x = o) = P\left(2 - \frac{a}{L}\right)$$

$$M_{\max} \begin{cases} \text{when } a < (2 - \sqrt{2})L = 0.586L \\ \text{under load 1 at } x = \frac{1}{2}\left(L - \frac{a}{2}\right) = \frac{P}{2L}\left(L - \frac{a}{2}\right)^2 \\ \text{when } a > (2 - \sqrt{2})L = 0.586L \\ \text{with one load at center of span} \quad = \frac{PL}{4} \\ \text{(Case 39)} \end{cases}$$

CASE 41 SIMPLE BEAM—TWO UNEQUAL CONCENTRATED, MOVING LOADS*

$$R_{1,\max} = V_{1,\max}(\text{at } x = o) = P_1 + P_2\frac{L - a}{L}$$

$$M_{\max} \begin{cases} \left[\text{under } P_1, \text{ at } x = \frac{1}{2}\left(L - \frac{P_2 a}{P_1 + P_2}\right)\right] = (P_1 + P_2)\frac{x^2}{L} \\ \left[\begin{array}{l}M_{\max} \text{ may occur with larger} \\ \text{load at center of span and other} \\ \text{load off span (Case 39)}\end{array}\right] = \frac{P_1 L}{4} \end{cases}$$

CASE 42 GENERAL RULES FOR SIMPLE BEAMS CARRYING CONCENTRATED, MOVING LOADS*

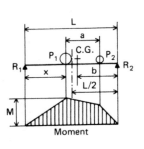

Moment

The maximum shear due to concentrated, moving loads occurs at one support when one of the loads is at that support. With several moving loads the location that will produce maximum shear must be determined by trial.

The maximum bending moment produced by concentrated, moving loads occurs under one of the loads when that load is as far from one support as the center of gravity of all the moving loads on the beam is from the other support.

In the accompanying diagram, the maximum bending moment occurs under load P_1 when $x = b$. It should also be noted that this condition occurs when the centerline of the span is midway between the center of gravity of loads and the nearest concentrated load.

Nomenclature: R = reaction, M = moment, V = shear, and Δ = deflection.
*From *Manual of Steel Construction*, 8th ed., American Institute of Steel Construction, 1980.

Appendix

B

Charts for Determining Stiffness Factors, Carryover Factors, and Fixed-End Moments for Beams with Variable Cross Sections*

---

*The charts in this appendix are taken from R. A. Caughey and R. S. Cebula, "Constants for Design of Continuous Girders with Abrupt Changes in Moments of Inertia," Iowa Engineering Experiment Station, Bulletin 176, vol. LII, Apr. 14, 1954.

200   Appendix B

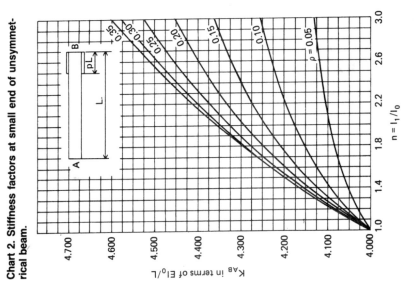

Chart 2. Stiffness factors at small end of unsymmetrical beam.

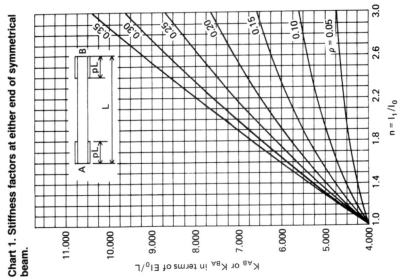

Chart 1. Stiffness factors at either end of symmetrical beam.

Stiffness Factors, Carryover Factors, and Fixed-End Moments **201**

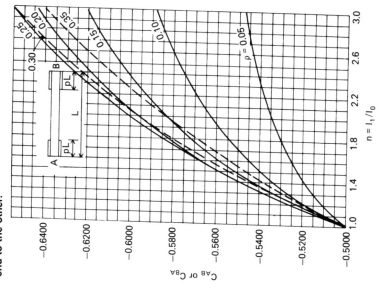

Chart 4. Carryover factors for symmetrical beam for either end to the other.

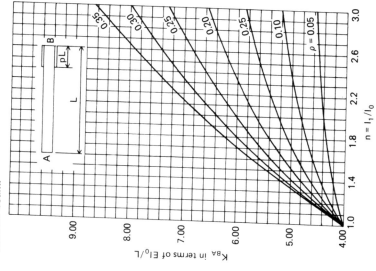

Chart 3. Stiffness factors at large end of unsymmetrical beam.

202 Appendix B

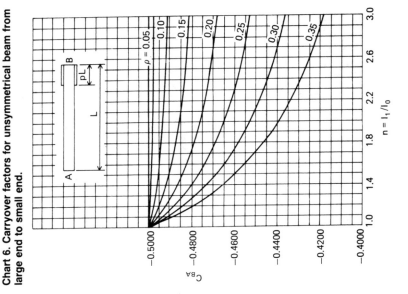

Chart 6. Carryover factors for unsymmetrical beam from large end to small end.

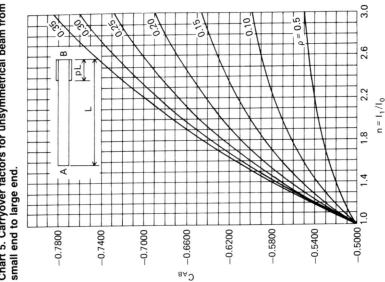

Chart 5. Carryover factors for unsymmetrical beam from small end to large end.

Stiffness Factors, Carryover Factors, and Fixed-End Moments **203**

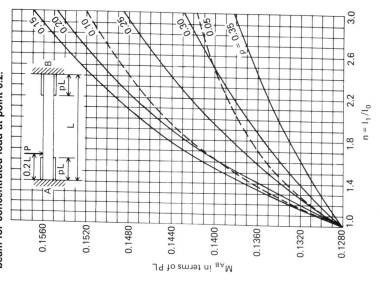

Chart 8. Fixed-end moments at left end of symmetrical beam for concentrated load at point 0.2.

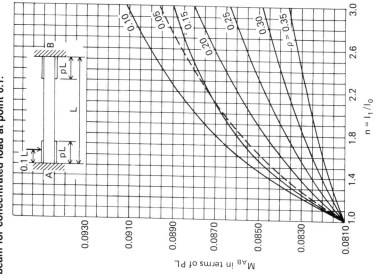

Chart 7. Fixed-end moments at left end of symmetrical beam for concentrated load at point 0.1.

204 Appendix B

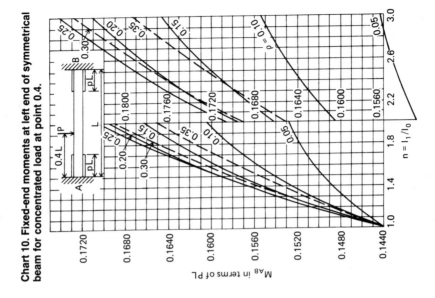

Chart 10. Fixed-end moments at left end of symmetrical beam for concentrated load at point 0.4.

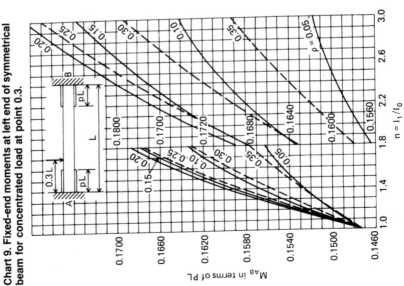

Chart 9. Fixed-end moments at left end of symmetrical beam for concentrated load at point 0.3.

Stiffness Factors, Carryover Factors, and Fixed-End Moments   **205**

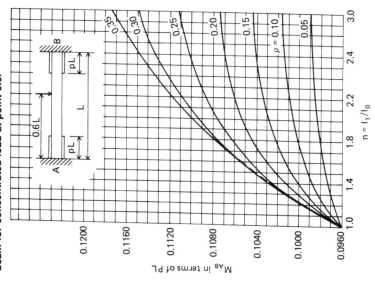

Chart 12. Fixed-end moments at left end of symmetrical beam for concentrated load at point 0.6.

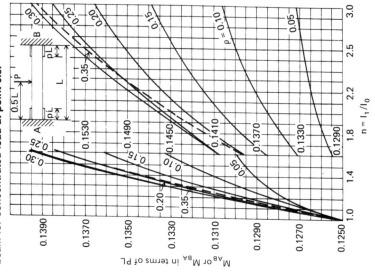

Chart 11. Fixed-end moments at left end of symmetrical beam for concentrated load at point 0.5.

Chart 13. Fixed-end moments at left end of symmetrical beam for concentrated load at point 0.7.

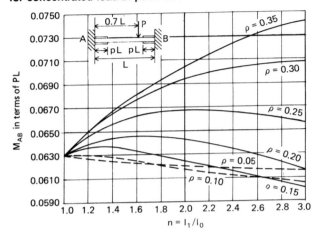

Chart 14. Fixed-end moments at left end of symmetrical beam for concentrated load at point 0.8.

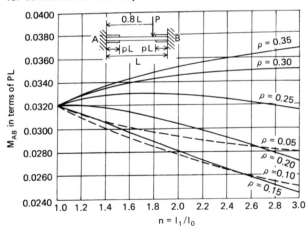

Chart 15. Fixed-end moments at left end of symmetrical beam for concentrated load at point 0.9.

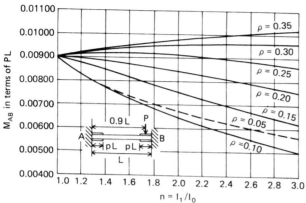

Chart 16. Fixed-end moments at large end of unsymmetrical beam for concentrated load at point 0.1.

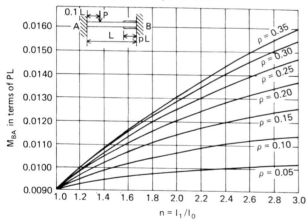

Chart 17. Fixed-end moments at large end of unsymmetrical beam for concentrated load at point 0.2.

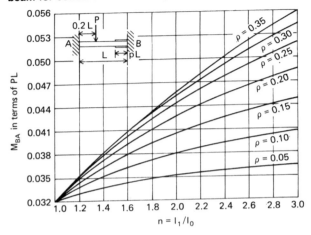

Chart 18. Fixed-end moments at large end of unsymmetrical beam for concentrated load at point 0.3.

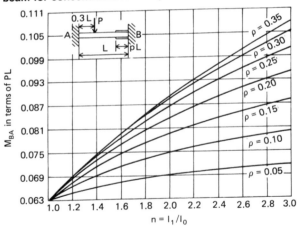

Chart 19. Fixed-end moments at large end of unsymmetrical beam for concentrated load at point 0.4.

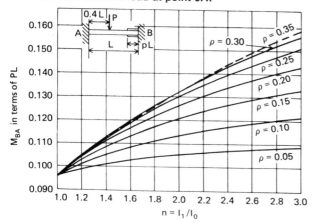

Chart 20. Fixed-end moments at large end of unsymmetrical beam for concentrated load at point 0.5.

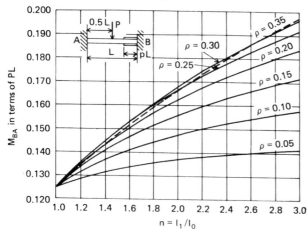

Chart 21. Fixed-end moments at large end of unsymmetrical beam for concentrated load at point 0.6.

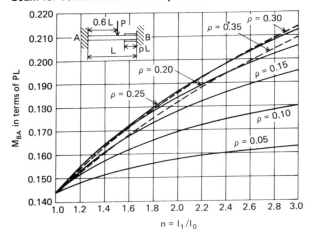

Chart 22. Fixed-end moments at large end of unsymmetrical beam for concentrated load at point 0.7.

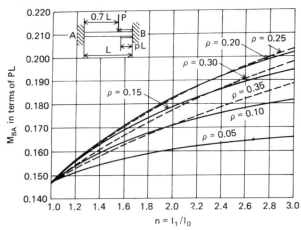

Chart 23. Fixed-end moments at large end of unsymmetrical beam for concentrated load at point 0.8.

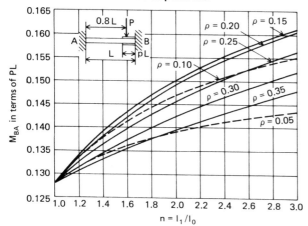

Chart 24. Fixed-end moments at large end of unsymmetrical beam for concentrated load at point 0.9.

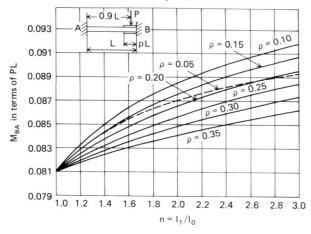

Chart 25. Fixed-end moments at small end of unsymmetrical beam for concentrated load at point 0.1.

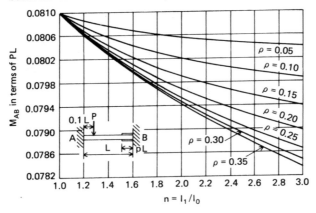

Chart 26. Fixed-end moments at small end of unsymmetrical beam for concentrated load at point 0.2.

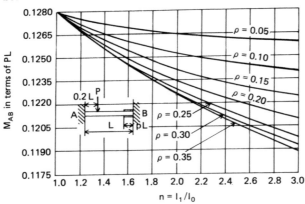

Chart 27. Fixed-end moments at small end of unsymmetrical beam for concentrated load at point 0.3.

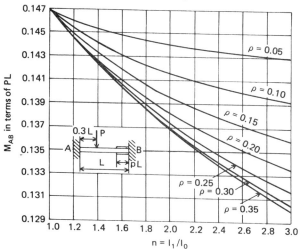

Chart 28. Fixed-end moments at small end of unsymmetrical beam for concentrated load at point 0.4.

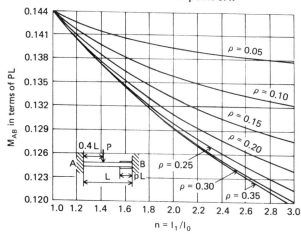

214 Appendix B

Chart 29. Fixed-end moments at small end of unsymmetrical beam for concentrated load at point 0.5.

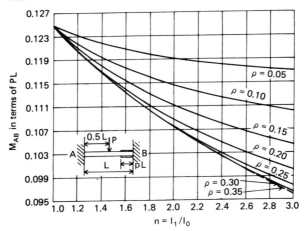

Chart 30. Fixed-end moments at small end of unsymmetrical beam for concentrated load at point 0.6.

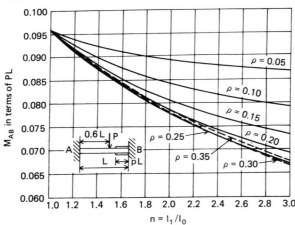

Chart 31. Fixed-end moments at small end of unsymmetrical beam for concentrated load at point 0.7.

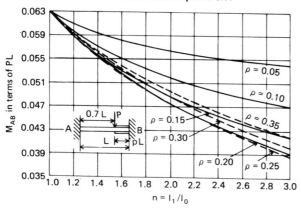

Chart 32. Fixed-end moments at small end of unsymmetrical beam for concentrated load at point 0.8.

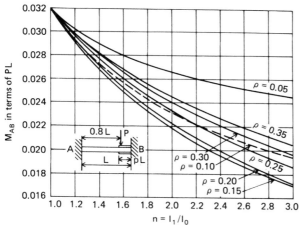

Chart 33. Fixed-end moments at small end of unsymmetrical beam for concentrated load at point 0.9.

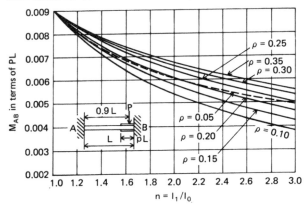

Chart 34. Fixed-end moments at large end of unsymmetrical beam for uniform load.

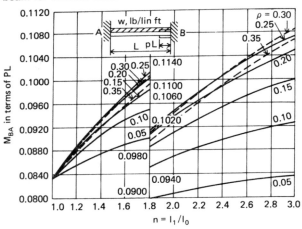

Chart 35. Fixed-end moments at either end of symmetrical beam for uniform load.

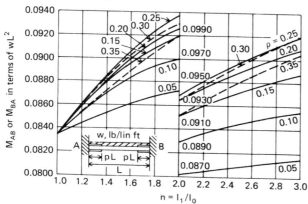

Chart 36. Fixed-end moments at small end of unsymmetrical beam for uniform load.

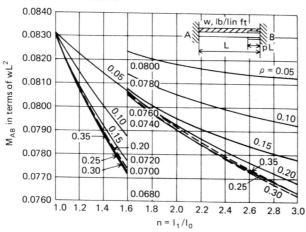

Appendix

C
Units of Measure and Conversion

TABLE OF CONVERSION FACTORS

Unit class	Multiply	By	To obtain
Length	Inches	2.54001	Centimeters
	Feet	30.4801	Centimeters
Area	Square inches	6.4516	Square centimeters
	Square feet	929.0359	Square centimeters
Volume	Cubic inches	16.3872	Cubic centimeters
	Cubic feet	28.317×10^3	Cubic centimeters
Weight*	Pounds, avoirdupois	0.4536	Kilograms
	Pounds per square foot	1.4882	Kilograms per square meter
	Pounds per cubic foot	16.0184	Kilograms per cubic meter
Stress*	Pounds per square inch	0.0703	Kilograms per square centimeter
	Pounds per square foot	4.8824	Kilograms per square meter
Moment*	Pound-inches	1.1521	Kilogram-centimeters
	Pound-feet	0.1382	Kilogram-meters
Tons	Tons, long	2,240	Pounds
	Tons, long	1,016.048	Kilograms
	Tons, long	$1,016.048 \times 10^{-3}$	Tons, metric
	Tons, short	2,000	Pounds
	Tons, short	907.1858	Kilograms
	Tons, short	907.1858×10^{-3}	Tons, metric
	Tons, long, per square inch	1.5748	Kilograms per square millimeter
	Tons, long, per square foot	10,936.592	Kilograms per square meter
Degrees	Degrees, angular	0.0174533	Radians
	Degrees Fahrenheit minus 32	0.5556	Degrees Celsius
	Degrees Celsius	1.799856, then add 32	Degrees, Fahrenheit

*If the unit is in kips, multiply each factor by 1,000 to obtain the metric system equivalent.

CONVERSION VALUES FOR MODULUS OF ELASTICITY E

Material	U.S. Customary System, kips per square inch	Metric system, kilograms per square centimeter
Steel	29,000 to 31,000	2,000,000 to 2,200,000
Concrete	2,000 to 6,000	140,000 to 420,000
Wood	1,350 to 1,700	95,000 to 120,000

UNITS USED IN DEFLECTION Δ FORMULAS

Equation	U.S. Customary System	Metric system
$\Delta = \dfrac{PL^3}{EI} C$ or $\Delta = \dfrac{wL^4}{EI} C$	P or wL in kips E in kips per square inch I in inches^4 L in feet × 12 inches Δ in inches	P or wL in kilograms E in kilograms per square centimeter I in centimeters^4 L in meters × 100 centimeters Δ in centimeters

Index

AASHTO (American Association of State Highway and Transportation Officers), 13
AISC (American Institute of Steel Construction, Inc.), 12
Angular rotation of beams, 37

Beams:
  angular rotation of, 37
  cantilever (*see* Cantilever beams)
  continuous (*see* Continuous beams)
  cover-plated (*see* Cover-plated beams)
  fixed at both ends, 50–58
  fixed at one end, 42–50
  fixed-end moments of, fundamental formulas for, 37–39
    table, 59–61
  floor, 12, 167
  influence lines of deflection of, 7, 90–91
  simple-support (*see* Simple-support beams)
Bridges:
  AASHTO specifications, deflection limit for, 13
  impact stress formula for, 169
  lane loading on, 168
  truck loading on, 168
Buildings:
  AISC specifications, deflection limit for floor beams, 12

Cambering, truss, 165
Cantilever beams:
  with constant cross section, 8–9
  cover-plated, 119–120

Cantilever beams (*Cont.*):
  deflection of: for concentrated load, 8
    for uniformly distributed load, 9
    with variable cross sections, 119–120
Carryover factor, 88, 149
Castigliano's theorem of least work, 6
Column-analogy method, 149
Conjugate-beam method:
  application of, for constant cross sections, 15–27,
    for variable cross sections, 119–137
  theorems of, 10–12
Consistent deflection method:
  application of, 85–86
  theorem of, 84
  trusses and, 175–178
Continuous beams:
  advantages of, 81
  analysis of, 81, 84, 87, 90
  consistent deflection method for solving, 84–86
  with constant cross section, 81–96
  deflection of (*see* Deflection of continuous beams)
  influence lines of deflection of, 90–91
  maximum deflection of, 96
  moment-distribution method for solving, 87–90
  support moments in, 81
  three-moment equation for solving (*see* Three-moment equation method)
  with variable cross sections, 147–164
Cover-plated beams:
  application of moment distribution for, 157–164
    factors in (*see* Moment-distribution method, application of, for variable cross sections)
  cantilever, 119–120

221

222 *Index*

Cover-plated beams (*Cont.*):
  continuous, 147–164
  deflection of (*see* Deflection of cover-plated beams)
  end moments and, 132–137
  fixed at both ends, 148–157
  fixed-end moments of, 147–148
  simple-support, 122–132
Cover plates:
  effect of, on deflection, 119–137
  effect of, on fixed-end moments, 147–157
Curve, elastic, 1, 7–9

Deflection limits, 12–13
Deflection of beams:
  cantilever (*see* Deflection of cantilever beams)
  computation of: by conjugate-beam method, 10–12
    by moment-area method, 7–10
  continuous (*see* Deflection of continuous beams)
  cover-plated beams (*see* Deflection of cover-plated beams)
  due to end moments, 40–41
  due to load and end moments, 58
  due to truck loading, 20–22
  effect of cover plates on, 119
  effect of end moments on, 37
  fixed at both ends: for concentrated load, 50–51
    table, 72–73
    for uniformly distributed load, 53–54
    table, 76–78
  fixed at one end: for concentrated load, 42–43
    table, 64–65
    for uniformly distributed load, 45–50
    table, 68–69
  influence lines of, 7, 90–91
  simple-support (*see* Deflection of simple-support beams)
  (*See also* Maximum deflection of beams)
Deflection of cantilever beams:
  for concentrated load, 8
  for uniformly distributed load, 9
Deflection of continuous beams:
  for concentrated load: two spans, table, 97–99
    three spans, table, 100–105

Deflection of continuous beams, for concentrated load (*Cont.*):
  four or more spans, table, 106–113
  for uniformly distributed load: two spans, table, 114
    three spans, table, 115
    four or more spans, table, 116–117
Deflection of cover-plated beams:
  for concentrated load: cantilever, 119
    continuous, 159
    fixed at both ends, 148–151, 154
    simple-support, 121–127
  for uniformly distributed load: cantilever, 120
    continuous, 161
    fixed at both ends, 152–153, 156
    simple-support, 128–132
  with end moments, 132–137
Deflection of simple-support beams:
  for concentrated load, 15–17
    table, 28–29
  for uniformly distributed load, 22–25
    table, 32–33
  due to truck loading, 20–22
Deflection of statically determinate structures, 6
Deflection of statically indeterminate structures, 81
Deflection of trusses, 165
Distribution factor, 89, 160

Elastic curve, 1, 7–9
Elastic limit, 1
Elasticity, modulus of, 1
End moments:
  cover-plated beams and, 132–137
  effect of, on deflection, 37, 40–41

Fixed-end moments:
  for constant cross section, 59–61
  fundamental formulas for, 37–39
    table, 59–61
  for variable cross sections, 147–148
  column-analogy method in computing, 149
  integration method for calculating, 147–148
Floor beams, 12, 167

Hooke's law, 1

Impact stress formula, 169
Inflection, point of, 42, 46, 51
Influence lines of deflection:
  of beams, 7, 90–91
  of trusses, 169–175
Integration method, 147–149

Lane loading, 168
Lateral stiffness:
  for constant cross section, 87
  for variable cross sections, 148–149

Maximum deflection of beams:
  for concentrated load, 17–18, 44–45, 52–53
    tables, 30–31, 66–67, 74–75
  for uniformly distributed load, 25–26, 48–50, 55–57
    tables, 34–35, 70–71, 78–79
Maximum deflection region:
  for concentrated load, 19
    table, 28–29
  for uniformly distributed load, 26
    table, 32–33
Maximum deflection of trusses, 167
  for truck loading, 168
Maxwell's law, 84–85
Modulus of elasticity, 1
Moment(s):
  fixed-end (see Fixed-end moments)
  shear force for, 39–40
  support, 81
Moment-area method:
  application of, for constant cross section, 8–10
  theorems of, 7–8
Moment-distribution method, 87–90
  application of, for constant cross section, 88–90
    carryover factor in, 88
    distribution factor in, 89
    fixed-end moments in, 37–39, 59–61
    stiffness factor in, 87
  application of, for variable cross sections, 157–164
    carryover factor in, 149

Moment-distribution method, application of, for variable cross sections (*Cont.*):
    distribution factor in, 160
    fixed-end moments in, 147–148
    stiffness factor in, 148–149
  theorem of, 87
Moment of inertia, 1
  tables, 2–6

Reciprocal deflection (Maxwell's law), 84–85
Redundant, 84–86
Rotation of beams, angular, 37
Rotational stiffness:
  for constant cross section, 87
  for variable cross sections, 148–149

Section moduli, table, 2–6
Shear force for moments, 39–40
Simple-support beams:
  with constant cross section, 15–27
    deflection of: for concentrated load, 15–17
      table, 28–29
    for uniformly distributed load, 22–25
      table, 32–33
    due to truck loading, 20–22
    maximum deflection of, 17–18, 25–26
      tables, 30–31, 34–35
    maximum deflection region for, 19, 26
      tables, 28–29, 32–33
  with variable cross sections, 122–132
Simpson's rule, 91
Slope of tangent, 81
Statically determinate structures, deflection of, 6
Statically indeterminate structures, deflection of, 81
Stiffness factor, 87, 148–149
Superposition method, 40, 169, 172–174
Support moments in continuous beams, 81

Three-moment equation method:
  application of, 82–84
  theorem of, 81–82
Truck loading, 168
  deflection of beams and, 20–22
Trussed beams, stress analysis of, 177–178
Trusses:
  cambering of, 165

Trusses (*Cont.*):
consistent deflection method and, 175–178
deflection of, 165, 167
determinate, 166
indeterminate, 175
influence lines of deflection of, 169–175
maximum deflection of, 167
superposition method and, 40, 169, 172–174

Trusses (*Cont.*):
truck loading and, 168
unit-load method and, 165–167

Unit-load method:
application of, 166–167
theorem of, 165

Young's modulus, 1

About the Author

Yun C. Ku is a consulting engineer with more than 30 years of experience in structural design and analysis involving bridges, dams, and industrial buildings. He has contributed significantly to many major construction projects throughout the United States. Prior to starting his consulting practice in New York State, Mr. Ku was an engineering editor for the Engineering Association of China and also published a book in China entitled *Flood Control Engineering*.